Taunton's

MASONRY COMPLETE

EXPERT ADVICE FROM START TO FINISH

The Taunton Press

The Taunton Press, Inc.,
63 South Main Street, PO Box 5506,
Newtown, CT 06470-5506
e-mail: tp@taunton.com

Editors: Joe Provey, John Ross
Copy editor: Katy Zirwes Scott
Indexer: Barbara Mortenson
Cover design: Kimberly Adis
Interior design: Kimberly Adis
Layout: Kerstin Fairbend
Illustrator: Christopher Mills
Photographer: John Ross
Front cover photographer: Scott Phillips

Library of Congress Cataloging-in-Publication Data

Macfie, Cody, 1976-
Tauton's masonry complete : expert advice from start to finish / by Cody Macfie ; editors of Fine homebuilding.
pages cm. -- (Taunton's complete)
Summary: "This volume is a comprehensive, authoritative, and easy-to-use reference for homeowners covering all home masonry projects from the most basic to advanced"-- Provided by publisher.
ISBN 978-1-60085-427-9 (pbk.)
1. Masonry--Amateurs' manuals. I. Fine homebuilding. II. Title. III. Title: Masonry complete.
TH5313.M28 2012
693'.1--dc23
2012028511

Printed in the United States of America
10 9 8 7 6 5 4 3 2 1

ACKNOWLEDGMENTS

I owe a special debt of gratitude to my wife, Carolyn, and our two boys, Jack and Thomas, for being patient enough to let me write this book amid our rigorous schedule and busy family life.

Thanks to John Ross, an outstanding person, editor, and photographer! Thanks also to Dan Morrison, for giving me my first shot at writing for *Fine Homebuilding* magazine, and to Chris Ermides, my former *Fine Homebuilding* editor, for bringing the possibility of this book to my attention. Special thanks go to Peter Chapman, executive editor, for promptly answering all of my questions and keeping us all on the right track. Thanks to the whole crew at Taunton, including Katy Scott, Christina Glennon, and Joe Provey.

Big thanks to Keenan Smith, for helping me make these projects a success, and to Sam Benton, for supplying us with the best material possible. A big hand goes to our hard-working masonry crews, who are ultimately responsible for creating these wonderful projects.

Thanks also to the homeowners and property owners for giving us a place to stage the projects in this book, and for allowing us to photograph the finished projects: Susan Van Ness, Keenan Smith, John Macfie, Chris Faulkner, Cemex, B & L Distributing Co., Inc. of S.C., Jack and Linda Lyons, 4-Rent Inc., French Broad Stone Supply, Inc., Vulcan Materials Co., Victor McHenry, Colin Shaw, Christ School, and Tom and Sebring Lewis.

—Cody Macfie

contents

DESIGNING WITH MASONRY 2

TOOLS AND TECHNIQUES 18

MATERIALS AND METHODS 50

PLANNING AND PREPARATION 70

DRY-STACK RETAINING WALL 82

STONE PATIO 98

>> >> >> >>

contents

BRICK VENEER

STEPS WITH WALLS

BLOCK WALL WITH STUCCO

BRICK WALKWAY

451

DESIGNING WITH MASONRY

CHOOSING WHAT TYPE OF MASONRY you will use on your project can be a challenge, especially since there are so many good options available. I live in a brick home that was built in 1923. There are Tudor-style stucco sections on the front of the house and a granite stone fireplace in the living room. Despite the variety, it looks great. You may live in a brick house, or a stone or stucco house, but you can still introduce other masonry treatments.

Stucco is a great finish to add to a retaining wall or as siding on a house. Stonework is attractive for just about everything, from garden pathways to front porches. Brick looks good when used in steps to the front porch or in a walkway leading to the front door. You can also mix masonry materials. Brick steps look just fine with stone treads. Keep in mind, however, that each masonry finish has its own unique appearance and installation process.

DESIGNING YOUR PROJECT

Gather design ideas well in advance of beginning your project. It's never too soon to start collecting pictures for ideas and inspiration. While there are plenty of examples in this book, the possibilities are really limitless. Look through magazines and other books to get an idea of what other people have done and what's popular. In this day of the ubiquitous phone camera, it's also possible to snap photos of treatments you like. Sometimes your own neighborhood is your best source of ideas, especially when the architecture of other houses is similar to yours.

Once you have an idea of what you like, make sketches to help you visualize the end results and draw up plans. Some projects will allow you to do some visualization right on the site. If you are building a walkway, for example, it is very helpful to mark the location with marking spray or a garden hose. Walk on the path several times to make sure it feels right. Now is the time to make any adjustments. If you are building a patio, you can use the same technique. Mark the patio's perimeter and actually use the location several times. If it will be for barbecues or a fire pit, go ahead and stage a cookout or build a fire. Invite some friends over to get their input while serving up some hotdogs and lemonade. Do whatever you need to be sure the plan is what you want. It will be hard to change after you lay brick or stone!

When building walls, take this process one step further. Once you have an idea of how you want the wall to look, build a mock-up at the site. Buy a small amount of stone and lay a small sample for the wall. This will give you a better idea of how the wall will fit into the landscape. If you're still unsure, build another mock-up with different stone or even brick. This may sound like a lot of work, but it will ensure that you are 100% satisfied with the results.

This short run of deep steps creates a relaxed and leisurely transition from the driveway to the entry path.

PATIOS

Patios are normally built adjacent to a house and serve as a transition to the yard. They can, however, be located in a secluded area on the property, such as a relaxing spot in the garden or a serene spot under an oak tree in the back corner of the yard. Some of my favorite island patios are nestled in a private space, with a fire pit in the center and a small sitting wall nearby. It is even better if there is a walkway leading to the area. This can be the perfect spot for social gatherings or an ideal place to eat lunch with your family on a warm summer day.

If you have a nice space in your yard but don't use it much, consider building a patio. Not only does a patio help define an area, but the hard surface created by bricks, pavers, or stone also creates a level place on which to set a table, chairs, hammock, grill, or screened enclosure. Patios can also be used as an entryway for a house, garage, outdoor kitchen, pool area, or garden shed.

➔ **For more on laying a stone patio, see p. 98.**

Patios are made of brick pavers or stone and can be dry-set or placed on concrete. Dry-setting pavers and bricks is easier and less time-consuming than setting them in mortar. It is also less expensive. The only disadvantage is that the surface won't be quite as smooth and might not be as permanent. Dry-setting requires shallow excavation, soil compaction, laying a crushed gravel base, more compaction, and setting stone or pavers on a bed of sand or fines (stone dust). Sand or stone dust is used to fill in the cracks when finished. One disadvantage with dry-setting is that the fines can eventually wash out during heavy rain, especially if the patio is on a slight angle. Another disadvantage is that you have to frequently sweep the patio to keep it clean. Polymeric sand is available to put between stones or brick pavers to prevent erosion and weeds from growing. It is similar to conventional sand but has water-activated chemical or organic binders that help keep particles in the joints, even during heavy rains.

Setting brick or flagstones in a mortar bed on a concrete slab is a little more expensive due to the cost of concrete, but it does make a patio more permanent. Stones set like this and bound by grout create a smoother, more secure surface that is easier to clean. There are no open joints if you build it right. This type of patio requires excavation, soil compaction, a gravel base, forming and pouring concrete, setting stones in a mortar bed, and grouting the joints. >> >> >>

This intimate backyard patio is nestled inside a curved retaining wall and sets the stage for the owners to enjoy a fire pit and a pretty view.

PATIOS (CONTINUED)

This entry porch, accented with matching arches, offers visitors shelter during inclement weather.

Covered patios create handsome transitions around a house's exterior.

A simple flagstone patio enhances this otherwise nondescript entry.

Patios can be finished in a variety of ways. This rustic flagstone has recessed grout joints where moss is encouraged to grow.

PATHS

Masonry materials are well suited to garden paths and walkways. Both flagstone and brick are proven path materials that are durable and relatively easy to install. Their surfaces are relatively smooth, and they are easy to conform to your existing site conditions.

There are several ways to build a walkway, each with its own level of difficulty. You can build a narrow path from stepping stones, a more substantial dry-laid path, or a path with mortared joints on a concrete base. Paths can be curved or straight.

A stepping-stone pathway can be very casual, intimate, and perfect for small, informal places such as gardens or woods. It is easy to build and generally consists of large stones set in a bed of sand or quarry stones. This type of path requires only limited excavation and little or no stonework.

Dry-laid walkways, with tightly fit stones or pavers, are a little more formal and require more work. They are usually made of brick or stone and are well suited for areas with high foot traffic, such as a pathway to your front porch. These walkways look best if they are 3 ft. to 5 ft. wide. They can be set on a concrete or a sand/gravel base. A dry-laid walkway requires some excavation, soil compaction, preparation of a gravel base, and a fair amount of attention to detail when setting the finishing materials. A final sweep of masonry sand, gravel fines, or polymeric sand in the cracks when you are done will give it a clean, finished look.

For more on laying a brick walkway, see p. 174.

A mortared sidewalk tends to produce the most formal appearance and requires the most work. Beyond preparing a concrete or thick-mortared bed, this type of construction requires the additional labor and expense of carefully fitting stones and grouting all the joints. However, a properly built mortared sidewalk will last the longest and is a highly valued feature for almost any landscape, whether you plan to sell or to keep your house for years.

Small details improve a path's appeal. Here, combining three short steps with a stone curb and path lights is superior to using just one step with no curb.

Brick and stone steps make it easy to navigate the slope of this approach.

PATHS (CONTINUED)

A straight flagstone path creates a more formal feel than a curved path.

The gentle curve of this flagstone walk is both inviting and visually appealing.

This flagstone path is a natural extension of the patio porch and is wide enough to handle a host of visitors.

Drawing Plans Drawings always make the job easier. If I have a client who is having a hard time with placement of a walkway, or who wants a good concept for the yard but can't make a decision, I recommend hiring a landscape architect. For my home, I asked a landscape architect to design a master plan for me. Five years later, my wife and I are still making the plan happen. You don't have to do it all at once, but it is smart to have a plan in place.

Some municipalities require you to have drawings and an approved plan before you begin. You can also purchase landscape plans online.

WALLS

The best thing about masonry walls is that they are highly functional and can be built almost anywhere. Masonry walls help prevent erosion, create privacy, provide places where people can gather, create visual and physical boundaries, or exist solely as accents. There are generally three types of masonry finishes for walls: stone, brick, and stucco.

➔ **See "Block Wall with Stucco," p. 156.**

The most common accent types are masonry veneers. They are laid on houses as brick, stone, or stucco to serve as siding. With proper fasteners, veneers can be attached to an existing wall or to a new wall, usually made out of wood sheathing, concrete, or block. Masonry walls can also be independent landscaping features, such as retaining walls or freestanding seat walls.

No matter what type of wall you build, you have the option of giving it a masonry finish. Consider using stone or brick that matches some other stonework nearby. Stucco is pretty neutral; it can harmonize with anything if you choose the right color.

Retaining walls

Retaining walls are used to hold back soil. They prevent erosion between two areas with elevation differences. Retaining walls built without mortar are called nonmortared dry-stack walls and are free of binding mortar and concrete footing. They should be made of large, flat stones or with masonry units designed for this purpose, such as keystone blocks, but not of brick or block. I have seen small retaining walls made out of small materials, such as small river stones or bricks, that fall down in short time. If a wall is not going to have cement, mortar, or steel reinforcement, it needs to be able to support itself and overcome the forces of nature.

See "Dry-Stack Retaining Wall," p. 82.

A mortared dry-stack retaining wall should be built on a below-grade footing. It requires drainage behind it to handle excess water and is backfilled with gravel. A good freestanding wall is built of large stones that are shaped to stack on top of each other securely.

» » »

RETAINING WALLS

The three basic types of retaining walls are a veneered wall with a concrete or block core, a solid mortared wall filled with mortar and rubble, and a nonmortared dry-stack wall. The height of the upper ground elevation will determine how high you need to make the wall. If it's higher than 3 ft., I recommend you build a mortared wall. A mortared wall allows the most freedom with regard to stone size, joint type, and height.

VENEERED RETAINING WALL

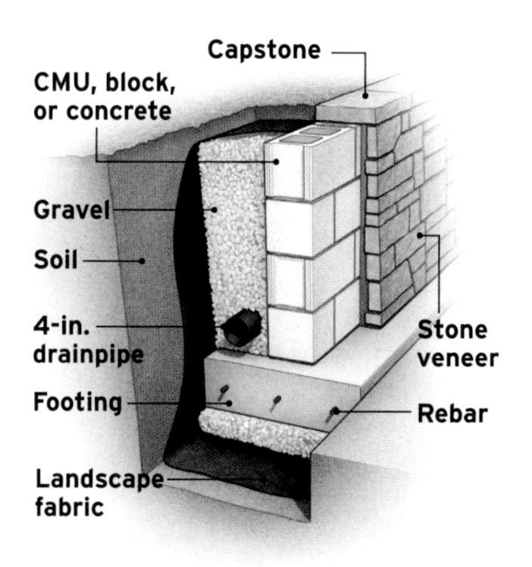

MORTAR AND RUBBLE-FILLED RETAINING WALL

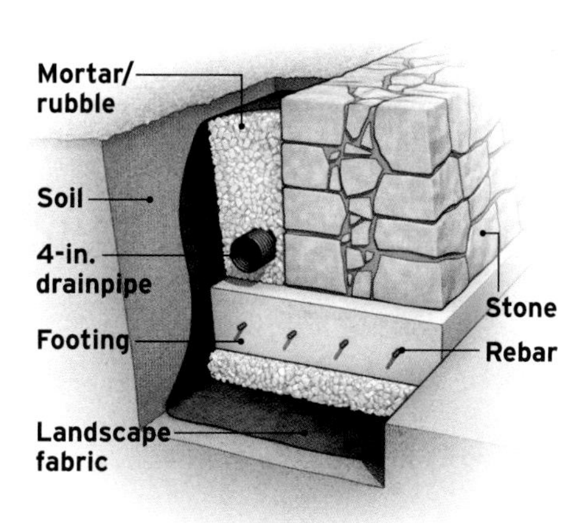

NONMORTARED DRY-STACK RETAINING WALL

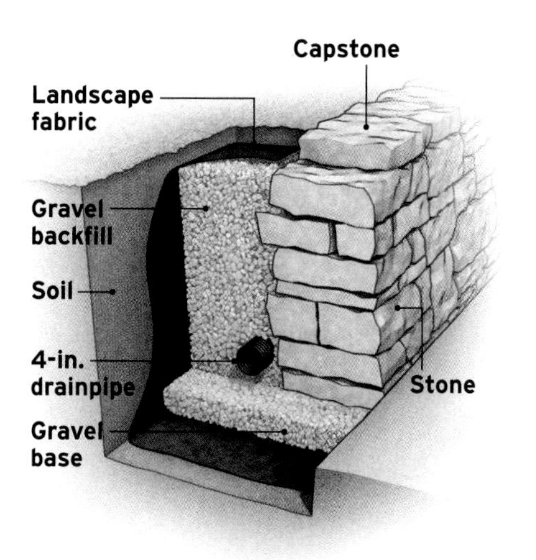

WALLS (CONTINUED)

Freestanding walls

While retaining walls can be built without mortar, freestanding walls typically need it. A nonmortared retaining wall leans into the soil it is retaining, making it stable. Freestanding walls, on the other hand, don't have this support. The pieces need to be mortared together and attached to a substantial footing. If you're building a brick or stucco wall, lay block construction first and veneer it with brick, stone, or stucco.

Freestanding walls have many uses, including seating, fencing, privacy, or camouflage: hiding unattractive residential mechanical units. I can visit just about any property and find a place for a wall. We recently built two long stone freestanding walls around a flagstone patio adjacent to a house. The homeowner wanted a place for guests to sit around a grill during social gatherings. The wall serves as a small seat wall during summer parties and separates the hard patio surface, used for eating, grilling, and sitting, from the outdoor lawn space used by the homeowner's children.

➔ **See "Fire Pit with Seat Walls," p. 220.**

If you want a freestanding wall detached from the house, you have a little more flexibility with design. Just because you have a brick house doesn't mean your wall needs to be brick. As I mentioned earlier, my brick home has a small stone fireplace in the living room. I built some small stone retaining walls in the landscape and used the same type stone and pattern that were already on the fireplace, and it looks great. A combination of brick and stone can look good in certain settings and the two materials set each other off attractively.

Using a variety of materials creates an effect greater than the sum of its parts. Here, the bold stonework and woodwork complement each other.

Rustic stonework veneer that is carefully shaped and applied using a dry-stack technique improves the appeal of an otherwise plain garage wall.

A simple stone wall turns an otherwise unsightly culvert end into a handsome landscape feature.

Construction techniques for building this fire pit and seat wall are no more complicated than for building a typical mortared wall.

The stone cap and stucco finish of this wall complement each other in both style and color.

COLUMNS AND STAIRS

Masonry columns are built using methods similar to those used for walls, but they perform supporting, rather than dividing, functions. For example, columns can serve as structural posts on a porch or arbor, gate supports, or mailbox posts. A short column can be used as a base for timber posts to give them a more substantial look. Columns are especially appropriate where other masonry finishes are nearby. The key to constructing any masonry column is to start with a solid core and to cover it with a masonry veneer (stone or brick alone is not load-bearing). The core can be veneered with stone or brick veneer, or stucco.

The complexity of steps and stairs varies greatly. Steps that connect two landscape elevations need be no more complicated than a few large stones set on a compacted base. On the other hand, stairs that lead up to an entry may involve multiple construction techniques and require highly accurate measurements and mathematical calculations.

Steps incorporated into a landscaped slope provide the most comfortable way to get from one elevation to another. They can be built with or without mortar. If you are not using mortar, choose steps that are large and flat for stability, because they will be supported only by the earth beneath them. Mortared steps require forms into which concrete is poured. You can then finish them with whatever masonry veneer you choose.

Masonry stairs (as opposed to landscape steps) require a concrete block base upon which a masonry veneer may be applied. If you have an existing set of masonry stairs, you can tear out the treads and risers and apply a new veneer of brick, stone, or stucco. If you have a set of wooden steps that you want to replace, you will need to demolish the wooden stairs, pour a footing, lay a concrete block base, and then veneer the blocks with a masonry finish. The risers can be finished with brick, stucco, or stone, but the treads should be made of brick or stone, because they are more durable. Choose a masonry finish that matches the surroundings. If you have a flagstone porch, consider using flagstone treads. If you have a brick house, consider using brick risers with brick or stone treads. If you have a stucco house, consider using stucco or coordinating stone risers with stone treads. Using a material that matches the existing structure will help the new elements blend with the old.

➔ **See "Steps with Walls," p. 134.**

>> >> >>

COLUMNS AND STAIRS (CONTINUED)

An object as mundane as a mailbox can become an attractive garden feature when encased in a stone column.

The rustic stonework of the column fits well with the rough stonework of the patio and provides a handy place to rest a morning cup of coffee.

A stone stairway provides safe access to the lower part of this yard and adds a handsome landscape feature.

This advanced project, stone steps with lighting, is a significant upgrade from pre-existing wood steps. (For more on this project, see p. 134).

LOCATION, LOCATION, LOCATION

Location is the key to making your project more enjoyable. You don't want to build a patio in an area that is going to be uncomfortable, such as a damp, shady section in the backyard. Consider a sunny spot. In addition, patios need to be scaled to fit the site. Just like the porch on the front of your house, a patio shouldn't be too big or too small. Also consider the terrain, soil, and proximity to trees. If you are building a retaining wall and need a footing, avoid placing it right in front of a 100-year-old healthy oak tree; the excavation would injure the tree's root system. Similarly, if you have a large sloping backyard and you want to build a patio, choose an area that is relatively flat. A slope will require building retaining walls to shore up and level the site.

The width of a walkway has a lot to do with its location on your property. I prefer a wide walkway if it leads to the front door or to a porch entryway, which is generally where most of the traffic flows. If you meet the mailman on the walkway, you don't want to have to step into the grass to let him pass. If you are having a party, you need enough width to accommodate constant foot traffic. A width of 4 ft. to 5 ft. is generally sufficient. Less formal walkways, such as garden paths, yard paths, or paths that lead to the back door, can be 2 ft. to 3 ft. wide. These are areas with less traffic (and are often places where you don't have as much room to work).

A retaining wall makes a great transition between two different elevations along a driveway.

STONEWORK

The best thing about masonry: It is timeless. Stonework never goes out of style and there are many different types from which to choose. By looking at different stonework options, you will probably be able to pick out what you like and dislike, or at least narrow it down to a few choices. Rough, dry-stack stonework looks great in informal settings, such as retaining walls and rustic homes. The stones for this style are left in their natural state, with little or no dressing (shaping). It is common to see old chimneys and foundations built in this style. In old homes, if the only type of rock available was very craggy, dense stone, the masons might have left the stones in their natural state to prevent having to spend a lot of labor dressing them.

Jointed stonework looks great in most settings. Each stone is stacked with a consistent mortar joint (typically about 1/2 in. to 1 in. wide) around it. There are several ways to finish the joints. You can recess them with a special tool, rake them flush with the stone faces, or leave them proud (raised above the surrounding material). With this type of stonework, you also have the option of using any color grout you want. Standard gray mortar is the most common choice, but I have done projects with white grout paired with dark gray stones, and brown grout with brown or tan stones. Jointed stonework looks great on just about any project, but especially on foundations and retaining walls.

When building a stone patio or walkway, flagstone is the most popular material to use. Flagstone can be set on a concrete base or on a bed of crushed gravel. Either way, the patio needs to have joints. Joints that are 3/8 in. to 1/2 in. wide and raked flush look best with flagstone on a concrete bed. For flagstone with a gravel bed, joints (up to 1 in.) are acceptable, though I prefer a narrow joint with this setting too. If you are planting a ground cover between the stones, consider leaving more space.

The interplay of the keystone, soldier stones, and the coursework offers interesting textural contrast to the rough-hewn timber that surrounds it.

This fireplace can afford to be bold, sited outdoors with lots of room.

A protruding rock creates a shelf upon which to set a drink or put a potted plant.

The two most common types of flagstone are pattern and irregular. Pattern flagstone is similar to tile. Each piece is already cut in squares or rectangles, or both, and is laid with all joints perpendicular. This type can be set on a mortared bed or on a gravel bed. Irregular flagstones are quarried and sold by the pallet in various shapes and sizes. You have to fit them together by using hammers, chisels, or saws to shape them accordingly. The more precise you are with the cuts, the more consistent the joints are—and the better the results. Irregular flagstones can be set on a concrete slab or a gravel bed.

This very rough stonework corner gives the impression that this wall has been here for generations.

Material Hunt If your project has a modest budget, you can hold off on the design details until you acquire the materials. Masonry materials are surprisingly easy to find at a discount if you are willing to look around. Planning well in advance can potentially save you money and make the process go more smoothly. Starting the process early gives you more time to shop for the best prices on materials.

Check with several material suppliers in your area to see if the material you need for your project is on sale. Also check newspapers or the Internet to see if anyone has material, such as rock or brick, for sale. Some people even give material away if you are willing to pick it up. For example, I recently bought some reclaimed paver bricks from my neighbor at a bargain-basement price and used them to build a 120-sq.-ft. patio. But before I found the brick, all I knew was that I wanted a patio behind my garage. I then let the found materials determine the style and size of the finished patio.

The careful stonework above this entry establishes a sense of order yet acknowledges the inherent randomness of stone.

Small details can make a big difference. Here, a mason created a stone front for an air vent to avoid interrupting the stone veneer.

BRICK

There are many reasons to use brick for your masonry project. Bricks insulate well, are attractive, and last a long time. If they go out of style or you get tired of them, you can paint them. You also have many colors, textures, and patterns with which to work, and you can choose colored grouts, as well.

The standard pattern is the running bond. It's clean, simple, and easy to install, and it works for any project. Running bond is used on most home veneers, as well as for walkways and patios. The bricks are set in horizontal rows, with each course staggered slightly to avoid "running" vertical joints.

The herringbone pattern is formal and has been around for a long time. It's a little more difficult to install because of all the end cuts that have to be made. You also have to measure angles when cutting, which takes more time. Herringbone pattern looks great on small patios, walkways, and fireplaces.

Basketweave is a simple, easy-to-install design. The bricks are laid horizontally and vertically in pairs, and are a nice alternative to running-bond patios and walkways.

Running-bond pattern.

Herringbone pattern.

Basketweave pattern.

The variety of grout colors to choose from is almost limitless.

STUCCO

Stucco is basically a decorative cover for visually unappealing surfaces. It is durable, attractive, and weather-resistant, making it suitable for many outdoor projects. Stucco is also less expensive than brick or stone. It requires lath, a scratch coat, and one or more finish coats made of portland cement, sand, and water. You can choose from a smooth finish, a trowel finish, a knockdown finish, a sand finish, and various pebble-dash finishes, to name just a few.

Stucco can be applied as a standard portland cement, sand, and water mixture and then painted whatever color you choose. Or you can purchase pre-mixed, ready-to-apply stucco in any color you choose. Allow standard stucco to cure for at least a month before painting it.

The beauty of masonry

While creating with masonry is a significant amount of work, the rewards are also substantial. Masonry has an inherent beauty and is, frankly, hard to mess up if you take your time. Also, masonry lasts a long time, so you will be enjoying what you produce for many years to come.

Whatever your design or project, this book will guide you through most basic masonry techniques. With simple, easy-to-follow instructions, you can build that masonry project you have always dreamed of. With masonry design, choose what you like and run with it. In masonry construction, start with the foundation, and go from there.

A stucco finish **is affordable, easy to apply, and gives predictable results.**

WORKABILITY

One factor that might help you decide which type of masonry to use is workability. Stucco is probably the easiest in terms of the materials and effort needed, and it does not take long to apply. Laying brick and block is a little more time-consuming, while stonework generally takes the longest because you have to be selective when choosing and setting each stone. In terms of material cost, stone is generally the most expensive.

TOOLS AND TECHNIQUES

MOST OF THE ITEMS NEEDED FOR a masonry project are not found in an average person's toolbox, but you won't have to look very far to find them. Home improvement and equipment rental stores carry many masonry tools. A basic collection will get you by, but consider specialized tools designed for specific masonry tasks, such as rock chisels and stucco trowels. They will make the job much easier and more satisfying. As a general rule, however, if you're not going to use a tool often and it can be rented, rent it. If it's something you'll use over and over, buy it.

Before you start your project, expand your masonry vocabulary with this list of tools and equipment.

BASIC TOOLS

EXCAVATION AND EARTH-MOVING TOOLS

DEMOLITION AND SPECIALTY TOOLS

CLEANUP TOOLS AND SAFETY GEAR

SHAPING TOOLS

Ever look at a stone structure and wonder how those stones were shaped to stack perfectly? The simple answer is that each stone was crafted with a hammer and chisel, the most important tools in masonry. You need them for a lot of jobs: to make stones smaller, shape squares, create corners, break stones in half, and make flat and curved faces. Aesthetically, these tools are essential to making your stonework look good. Structurally, they will help you shape stones correctly to create a strong project. With a little practice, in a short time you'll be able to use a hammer and chisel to shape any stone.

Hammers

Hammers for stonework include a brick hammer for refining the edges of stones, a rock hammer for shaping rocks, and a sledgehammer for breaking large rocks. Brick hammers get the most work on my crews and are used for light shaping and edge work, which is what they are designed for. However, a skilled mason will often use brick hammers to score cut lines, pry joints, set fasteners, secure twine, and set stones in mortar. Brick hammers are available in different weights, but I prefer 22-oz. hammers because they are light enough to use for long periods and heavy enough to shape stones effectively.

Rock hammers and sledges are primarily used to break rocks. A heavier sledge, with a 12-lb. head, will make short work of breaking large fieldstones into smaller pieces. Smaller sledges of 3 lb. or 4 lb., when used with chisels, are great for shaping bulky stones.

A 22-oz. brick hammer is a versatile tool to own—and can be used for more than just breaking bricks.

A small sledge and chisel are essential for shaping bulky wall stones.

Rock hammers with a wedged end for breaking stone are more specialized than typical sledges.

Chisels

Chisels are used for scoring, edging, refining, and breaking brick, rock, and block. Stonemason's chisels are heavy-duty with a thick shank and broad head. The better ones have replaceable carbide tips that hold an edge longer.

Chisels are sold as either handset chisels or hand-tracer chisels. The former has a thicker edge and is used to break off larger pieces of stone. The latter has a sharper tip and is used to remove smaller pieces of stone or to split a stone along a straight line.

Cold chisels and brick chisels are used for scoring brick and block. They do not have carbide tips and will not last nearly as long as a carbide-tipped chisel. However, they can be found at most home improvement stores and are generally less expensive. A cold chisel is a chisel made of tempered steel. Both carbide and cold chisels can be sharpened on the job site.

See "Sharpening Tools on the Job," p. 25.

Chisels for masonry are stout with thick shanks and wide heads. This shank's mushroomed end is typical of a well-used chisel.

The hand-tracer chisel (left) has a sharper edge than the handset chisel (right).

Handset chisels are used to remove large chunks from bulky wall stones.

A hand-tracer chisel, with its sharp tip, can make quick work of breaking stones.

WHAT'S IN THE BUCKET?

Every mason carries personal tools to the job (usually in a 5-gal. bucket). The tools listed here will simply be referred to as "masonry tools" in the "What You'll Need" lists that accompany the projects in this book.

- Brick hammer
- 3-lb. rock hammer
- Rubber mallet
- 2-in. handset chisel
- 2-in. hand-tracer chisel
- Mortar trowel
- Pointing trowel
- Whisk broom
- Tape measure
- 4-ft. mason level
- Roll of nylon mason twine
- Chalk line
- Line level
- Plumb bob

TRADE SECRET

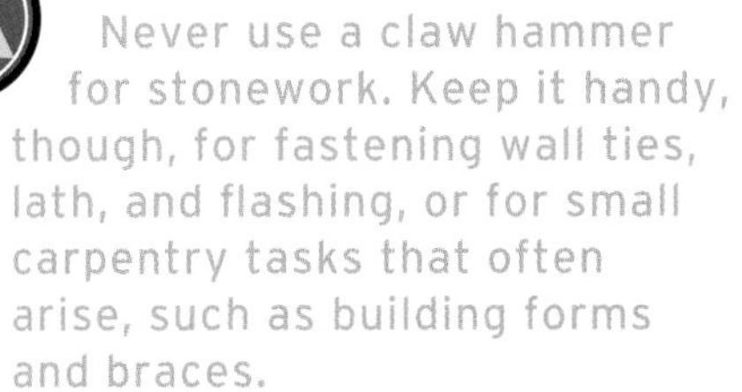

Never use a claw hammer for stonework. Keep it handy, though, for fastening wall ties, lath, and flashing, or for small carpentry tasks that often arise, such as building forms and braces.

TEN COMMON USES FOR A BRICK HAMMER

A brick hammer's versatility makes it a valuable tool on a job. Here are a few of its most common uses.

❶ **Remove large sections to shape a fieldstone into a brick shape.**

❷ **Set stones in mortar with either the head (as shown) or the handle butt.**

❸ **Break manufactured stone to length with a sharp rap of the pointed end.**

❹ **Dress the front edge of a fieldstone cap, in this case to a slight curve for a fire pit.**

❺ Cleanly break concrete block by cutting a groove and then striking it sharply.

❻ Score a break line on a large piece of flagstone and break the stone with a single strike.

❼ Dress a flagstone edge, after it has been broken, to remove irregularities.

❽ Break a solid brick in half. Do this on a bed of sand to cushion the brick so that the break will be more accurate.

❾ Adjust joints by prying the sharp end of the hammer between stones. This works well for small adjustments.

❿ Mark a cut line on a stone. Using the same tools in efficient ways reduces wasteful effort (like picking up a pencil to mark a cut line) throughout the day.

MAKE A CLEAN BREAK USING CHISEL AND SLEDGE

Place the chisel at the beginning of the cut, angled at about 30 degrees ❶. Then strike the chisel medium hard (less than all you've got but hard enough to make the stone take notice). Hit the stone only once. Move the chisel forward along the break line, reverse the angle, and strike the chisel, again, medium hard ❷. Proceed across the stone, reversing the chisel angle each time. When you reach the other side, start back across the stone along the same break line ❸. Keep doing this until the stone breaks ❹. It may take a few minutes; be patient, and you'll be rewarded with a clean cut.

1 Angle the chisel 30 degrees to one side and strike it once to begin the breaking process.

2 Move a couple of inches along the desired break line and reverse the chisel angle before striking the chisel again.

3 Advance across the stone, striking once in each location and alternating chisel angles.

4 Retreat along the break line in the same manner until the stone breaks.

DRESS A CORNER STONE WITH HAMMER AND CHISEL

To dress a flagstone corner for a patio, find a piece of stone that looks like a natural corner stone with an angle close to 90 degrees ❶. Use a framing square to mark a 90-degree angle on the edge with a pencil. Next, use a handset chisel and small sledge-hammer to remove small sections along the line ❷. Break off small pieces, working away from the edge ❸.

SHARPENING TOOLS ON THE JOB

Brick hammers, cold chisels, and carbide-tipped chisels can be sharpened on the job site with a portable angle grinder. Using a fine-grit wheel, gently pass the grinder across the tool's cutting surface. Try to match the existing bevel on the blade and avoid being too aggressive with the grinder. Keep a bucket of water handy to cool the steel often (photo at right below). You don't want the edge to get hot and lose its temper (ability to keep an edge).

1 Select a stone that already has a corner close to 90 degrees, or create one by breaking the stone perpendicular to its straight edge.

2 Remove irregularities along the edge that will be most visible.

3 Work away from the edge and be careful near the corner, where stones break easily.

SETTING TOOLS

Setting tools help you spread and pack mortar around stones. Trowels and pointers are considered setting tools. A rubber mallet is also a setting tool used to tap or set a stone into mortar without the risk of marring its surface or breaking it in half (which a sledgehammer might do).

In practice, a mason might consider tape measures, short levels, a builder's square, chalk lines, and marking pencils as setting tools because they reside in the same bucket and are used to set stones.

There are no hard and fast rules about which tool to use when you are setting stone. Often it's the tool in your hand that will get the job done fastest.

Trowels

A mason's trowel (sometimes called a brick trowel) has a diamond-shaped blade and is used for working with both wet and dry mortar. It is the mainstay of mortar work. A 10-in. mason trowel is good for working with block, brick, and stone. A large mason trowel also makes a great dustpan and brick setter. Trowels are usually available in 10-in. to 13-in. sizes. I have a 4-in. trowel the same shape (called a pointing trowel) that I use for laying brick and small stonework in tight spaces. The shape and size of the blade are ideal for transferring small amounts of mortar and spreading small patches. One person on my crew ground down his pointing-trowel blade to 3 in. and uses it for digging mortar from bucket bottoms and packing mortar between wall stones.

See "Spreading Mortar for Concrete Block," p. 29.

A tuck-pointing trowel has a narrow, flexible blade. It is primarily used to set mortar into joints. Tuck-pointing trowels are also used for removing mortar from between stones to leave sufficient room for grout, and for packing and shaping mortar when grouting. These trowels are available with blades of different widths. I generally use a 1/4-in. to 3/8-in. one, but you can get them as large as 3/4 in. for wider joints.

See "Dry-Grouting Flagstones and Capstones," pp. 30–31.

Large brick trowels are preferred for setting block because they pick up more mortar.

No tool a mason uses has only one function. The heft of a large brick trowel makes it great for tapping bricks and stones into place.

Buckets are extremely useful on a masonry job for toting and mixing mortar; this brick trowel with a ground-down point makes it easier to access mortar at the bottom of the bucket.

A good finishing trowel, used to smooth concrete, has good balance and solid construction.

Tuck-pointing trowels are excellent for packing grout between capstones and for shaping joints.

Metal stucco trowels can be used for applying brown coats and stucco.

Margin trowels are excellent for working in confined areas or when grouting wide joints, as shown here.

Other common trowels include finishing trowels for spreading brown coat and smoothing concrete, plastic finishing trowels for applying stucco, and margin trowels for working wide mortar joints. Stucco trowels are your best option for applying a scratch coat, or brown coat or for smoothing small areas of concrete, such as footings. They are available in 11-in. to 14-in. lengths and 4-in. to 5-in. widths. Other trowels used for general masonry include edgers, floats, and corner tools.

No matter which trowel you use, keep it clean and it will last longer. Once mortar sets up, it will be much harder to clean off. Over time, mortar buildup will weaken the bond where the handle meets the blade, eventually causing it to break off. Clean your trowel several times a day when possible.

POINTERS AND JOINTERS

Pointing and jointing tools are primarily used to finish joints between brick, block, and stone. Pointers are used to pack mortar into the joints of brick and stone. Jointers are used to finish the joints of stone, brick, and block work. They vary in shape, size, length, and material. Bullhorn jointers, concave stone beaders, hard jointers, double-blade slickers, wheeled-joint rakers, brick jointers, grapevine jointers, and stone and brick beaders are just some of the names you'll encounter. Most of these are for brick joints, but some apply to stone and block.

For some stone joints, such as the extruded joint, it is essential to buy a jointer for this purpose. If your block wall is going to be veneered with brick or rock, or covered with stucco, it doesn't really matter how you finish the joints. Your local masonry supply store should have samples of various joints and the tools with which to create them.

A T-shaped pointing trowel creates a grapevine joint profile.

Each of these pointing trowels leaves a distinct profile. Profiles from left to right will produce a recessed V shape (grapevine jointer), a raised curve (brick jointer), a groove (grapevine jointer), and a recessed curve (convex jointer).

Jointers can also be homemade. This jointer leaves an extended joint and is made from a $1/2$-in.-dia. piece of PVC pipe.

SPREADING MORTAR FOR CONCRETE BLOCK

There are a couple of trowel skills you'll need when spreading mortar for block work: spreading mortar for bedding joints on a footing or on the top of each course, and buttering the block ends. Keeping the mortar on the trowel will be the challenge because it has a tendency to slide off, but you'll get the hang of it with a little practice.

To spread mortar on a footing, put a little on a trowel and flip the trowel down with your wrist, about 3 in. to 4 in., and come to an abrupt stop. This will level the mortar on the trowel ❶. Rotate your wrist sideways and make a downward slicing motion with the trowel. The mortar should spread in a straight line on the footing ❷.

To butter ends or spread mortar on top of blocks and bricks, use the same technique to flatten the mortar on the trowel. Then turn the trowel upside down and scrape it sideways down the edge of the block. You have to do this rather quickly so that the mortar doesn't slide off the trowel before it gets to the block ❸. Do the same if you are spreading mortar on the top of a block ❹.

To adjust and set blocks, tap lightly with the trowel blade or handle ❺. Avoid pushing on the block, because it is easy to overadjust. Scrape away excess mortar by angling the trowel and slicing upward ❻. This will remove the mortar cleanly without smearing it on the block.

1 Remove excess mortar from a trowel by stopping its downward motion abruptly.

2 Spread mortar on a footing by using a cake-slicing motion.

3 Use the same motion to butter block ends before placing them.

4 Throw mortar on top of a block with a slicing motion, too, but let the trowel blade pass the edge of the block.

5 Gentle tapping is all that's needed to set concrete blocks in properly mixed mortar.

6 Remove excess mortar by slicing it off with a trowel and pulling the blade away. Avoid smearing the mortar.

DRY-GROUTING FLAGSTONES AND CAPSTONES

There are a few advantages to dry grouting over wet grouting. It doesn't take as long to apply, and it is a cleaner process. The disadvantage is that when the mortar dries, it is not as dense as wet grout because it's not mixed with as much water. Water can therefore penetrate the joint over time. I recommend dry grouting on porches or indoors, places that are not as exposed to water.

Mix the mortar and sand using the recipe on p. 61. Add just enough water to the mix so it retains its shape when you form a ball in your hand ❶. If it crumbles, add more water. If it sticks to your hand or glove, you've added too much water.

Pick up some mortar with your trowel and press it into the joints with a tuck pointer ❷. Press firmly and pack enough so that it rises above the flagstone surface 1/4 in. ❸. Let the mortar dry slightly and then rake it flush with the top of the flagstones using the blade of your tuck pointer ❹. If the mortar leaves stains on the edges of the stones, let it dry for a little longer and then try scraping it off. Do not let the mortar dry for too long or it will be impossible to remove.

As you work, scrape the excess into piles. Let the joints dry for about 30 minutes and then sweep away the dust and debris ❺. If there is staining on the stones, use a nylon brush to scrape it off. Do not put too much pressure on the joints until after they cure for a day or two.

1 **Mix joint mortar with enough water so that it hangs together when pressed into a ball.**

4 **Scrape away excess mortar once it's dry enough to come away without staining the stone's surface.**

2 Fill joints using a brick trowel to hold the mortar and a tuck pointer to press it into the joints.

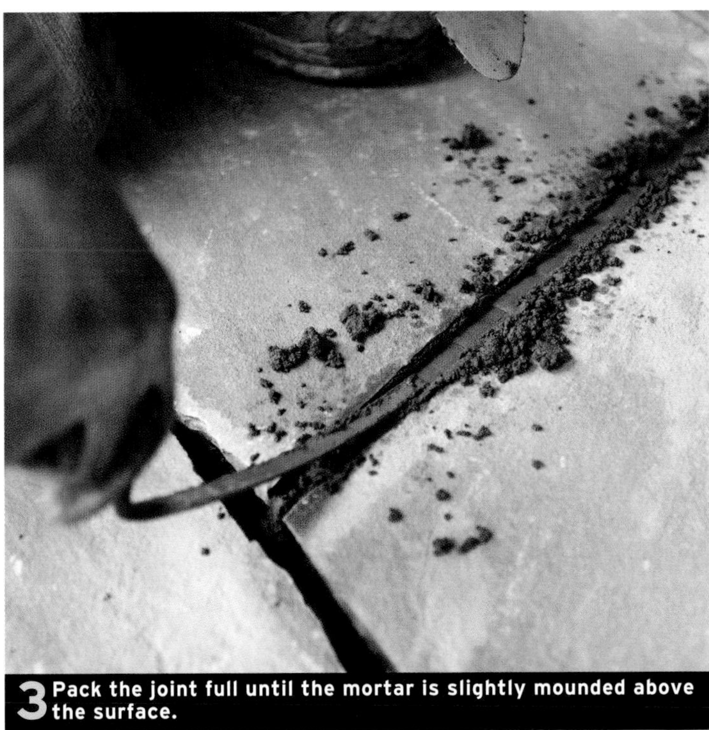

3 Pack the joint full until the mortar is slightly mounded above the surface.

5 Remove mortar dust by sweeping it into a dustpan.

LAYOUT TOOLS

Layout tools are a mason's secret weapon. They are the tools that make the job easier and allow the mason to deliver perfect, accurate work as opposed to mediocre work. Sometimes these tools cost a little extra, but the time and money spent to make the work precise will pay dividends in the end.

Generally, a layout tool is anything that helps you decide where to put the stones, brick, or blocks according to the plan. Measuring and marking tools include tape measures and marking sprays, masonry twine, garden hoses, and plumb bobs. Levels, lasers, and transits are also important layout tools.

Measuring and marking tools

A quality tape measure is essential. Masonry work is hard on tape measures, so it's worth spending a little extra to buy one that will last. You'll use a tape measure for all the phases of masonry work, from planning and layout to setting capstones.

There are other measuring tools that are not essential but if you are planning on doing a lot of masonry work, you might want to invest in them or at least learn how they work. A measuring wheel (or walking tape) measures as you walk along rolling its wheel. This tool is great for measuring long walls and long brick walkways to determine how much material you will need. For more accurate measurements of long walls or walkways, an open-reel tape works great and is readily available in 50-ft. and 100-ft. lengths. A brick mason ruler is specifically designed to measure brick coursing in increments relative to brick sizes.

Stringlines are a necessity when laying stone corners. Nylon twine works particularly well because it allows you to stretch it tightly in a straight line. The twine is also easy to tie off at the ends. When laying stone to a stringline, stay slightly to the inside of the twine so that it stays straight. Stringlines are also handy when laying courses of block and brick, establishing wall heights, and using batter boards (see p. 34).

Because of its flexibility, a garden hose is a perfect tool for laying out a retaining wall or benches around a fire pit (see the bottom left photo on the facing page). A rope or even a row of bricks will work almost as well. Once a layout is determined, use marking spray to mark it and remove the garden hose or whatever else was used to mark the outlines. >> >> >>

Good tools improve your work when they are used correctly. Here, the author checks a retaining wall for the correct batter.

A durable tape measure may cost twice as much as a lesser brand, but it will last three times as long.

When pulled tautly, nylon mason twine is as straight as the most expensive level and can be used to align stone in almost any situation.

Set stones as close to the stringline as possible—but without touching it. Otherwise, your reference line will be compromised.

A common garden hose is a useful tool for establishing the contours of a project layout.

The nozzle of a marking spray can is designed to allow you to point the can straight down while you spray.

LAYOUT TOOLS (CONTINUED)

USING BATTER BOARDS

Batter boards are horizontal boards nailed to posts placed just outside the corners of a project (which might be the construction of a new home or the building of a small patio or deck). They provide an easy way to establish the stringlines used for layout. For example, batter boards and stringlines can establish the desired height of a retaining wall, the location of a footing, or the outline for an excavation. Once stringlines have been established, batter boards are marked to preserve the string locations. The strings can be removed during excavation and construction but replaced later to continue serving as a reference.

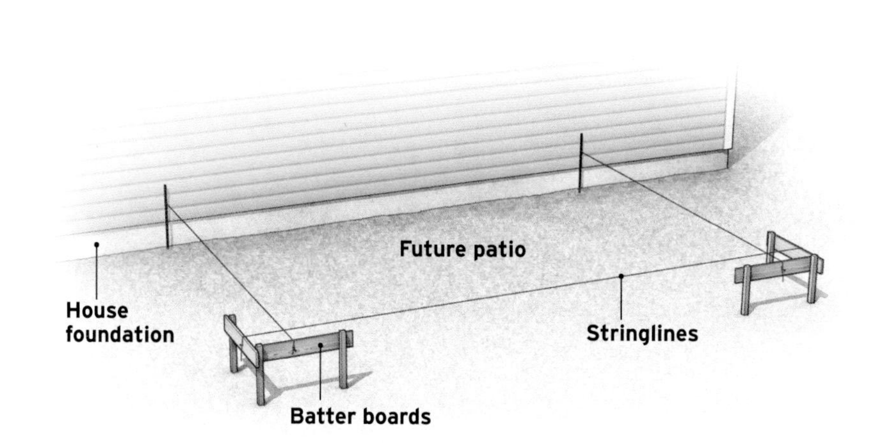

Levels, lasers, and transits

Levels are essential tools for masonry and come in various sizes, from 4-in. line to mason levels 6 ft. or longer. Whether you are laying block, brick, or stone, you need to make sure everything is level, including caps, footings, and coursework. Use a level often because laying one stone out of level on a stone wall can visually throw off the entire pattern. It can also set the stage for many unlevel stones to come.

I keep a 10-in. torpedo level with me to check my work in tight spots. However, torpedo levels allow you to check only one unit at a time, so keep a 4-ft. level close by for checking several stones at once.

A 4-ft. level is standard and the size I use most often. It is also great for determining the pitch of a concrete slab before laying flagstone. A 6-ft. level is handy for projects such as patios and walkways, where it's necessary to align stone tops and edges. It can also be used to flatten bedding gravel and align brick coursework.

Levels are also made of different materials, including wood, plastic, and aluminum. I have a 4-ft. wooden level that I use all the time, but it requires an occasional application of oil to preserve the wood. A wood level is not more accurate than an aluminum level per se, but I like the aesthetic of wooden levels; that, for me, is good reason to pay more to own one. For masonry, most levels are accurate enough but they will take considerable abuse, so it's wise to buy a sturdy level. Many top-of-the-line levels come with protective cases. Whatever level you buy, be sure to wipe off the cement frequently and avoid hitting it harder than a tap or dropping it on the ground.

Laser levels and transits are primarily used by surveyors and site preparers, but they come in handy for small masonry projects, too. A laser level sits on a tripod and can spin 360 degrees, projecting a red laser beam across a horizontal plane. A transit also mounts on a tripod, but instead of casting a laser beam, you sight the elevation reference you need. Both will help you determine distant elevation changes. We primarily use a laser level to determine elevation changes for retaining walls and patios. >> >> >>

BUILDING A TAPERED COLUMN

Building a tapered column is almost as easy as building a plumb column. For the column shown here, the taper was 4 in. from the bottom to the top as measured from the faces. When setting your stringline, mark the perimeter of the column on the ceiling (or as in this case, on the framing). At each corner, drop a plumb bob to the floor and mark the point. Connect the marks to form a square. Then measure out (4 in. in this case) to mark the column's base. Run stringlines from the perimeter corners at the ceiling to the corners of the base.

Torpedo levels are handy in tight spaces, so keep one in your pocket. For greater accuracy, however, use the longest level the space allows.

A 4-ft. level is the most versatile leveling tool for masonry. It's long enough to level courses of brick, block, and stone, yet short enough to level a narrow walkway or patio.

A 6-ft. level is as useful as a straightedge as it is a level. Its one drawback is that it is prone to getting bent, especially on a rough-and-tumble masonry jobsite.

A transit sits atop a tripod and has a lens through which to sight a reference point; a laser level has no lens but emits a light beam.

PLUMB VS. LEVEL

Plumb means exactly vertical in two planes. A plumb bob, used by masons and carpenters to establish a plumb line, is a pointed weight attached to a string.

Level can refer to either a horizontal line, such as the leading edge of a wall, or to a two-dimensional horizontal surface. In the case of the latter, the surface is level only when it is perfectly horizontal in two directions. A level is a tool that consists of a straightedge in which there is an encased, liquid-filled tube that contains an air bubble. When a level rests on a surface, the surface is said to be level only when the bubble is in the center of the tube. Depending upon the orientation of the tube, the level may be used to check for levelness or for verticalness. By checking a post or wall for vertical on at least two faces, a level may be used to check for plumb.

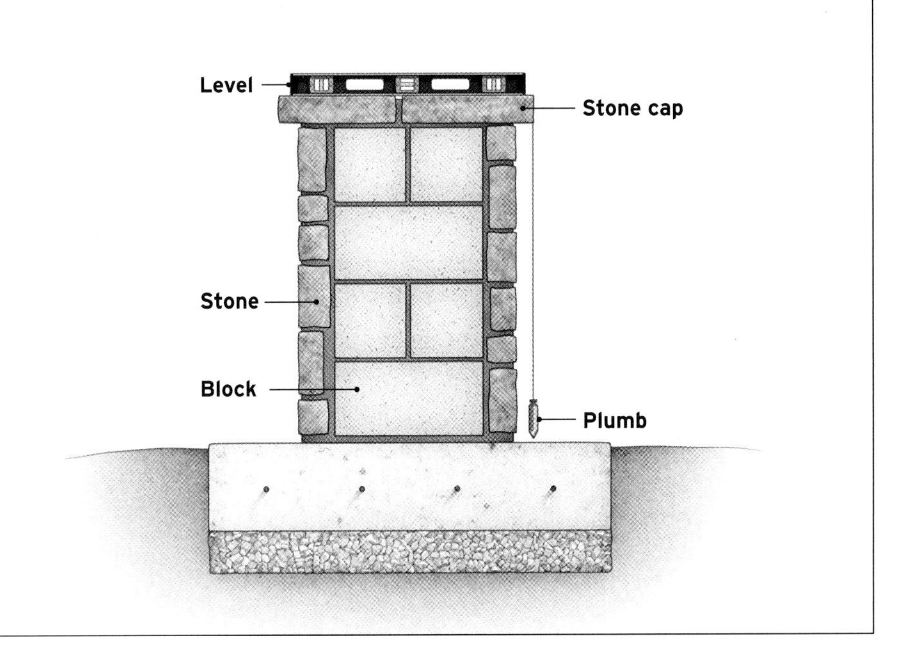

SEVEN WAYS TO USE STRINGLINES

Setting and using stringlines is integral to masonry. The more comfortable you are using stringlines, the faster and better you will work. On the job, stringlines directly reference a stone's edge as when setting bricks, columns, and patio edge stones. Stringlines can also be used as general reference to set stone elevations and wall thicknesses, and to form locations.

❶ Set brick and block coursework level and to the correct height.

❷ Build plumb column corners. For best results, accurately cut and fasten a plywood top to which stringlines can be attached.

❸ Align the level edge of stair treads (or the edge of a patio).

❹ Create a grid from which you can set the height of patio stones.

❺ Establish the location and dimensions of a wall. Even the roughest field wall will look better set to a stringline.

❻ Tie to batter boards to provide guides throughout the construction process.

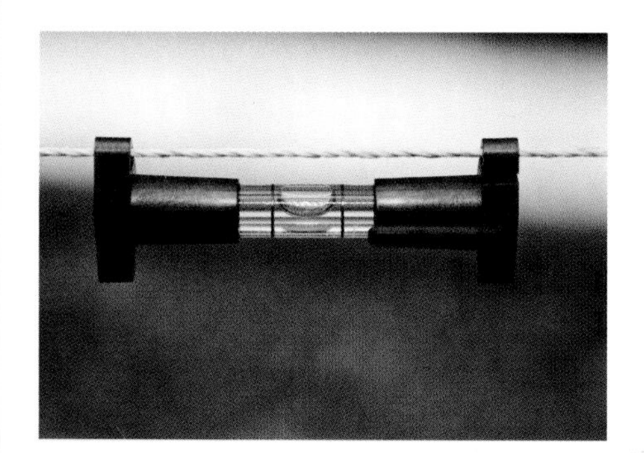

❼ Transform twine and a line level into a long level. For best results, place the line level in the middle of the span.

The easiest way to secure the stringline is to pass the loose end under the taut line.

Using a transit when installing a drain line

When installing a drain line, the easiest and most accurate way to make sure you have the slope you need for drainage is to use a transit. First, set up the tripod outside the work zone so you can view the entire project when panning across the area ❶. Adjust the tripod to the desired height and adjust the transit on the tripod to a level reading ❷; failing to do so will compromise your readings. Have a helper hold an extended tape measure on the slope at the drain-line location. You will be able to read the tape measure through the lens ❸. Move the tape to several locations. The longest (highest) reading will indicate the lowest spot.

1 Set the tripod on a stable surface outside the work zone.

2 Level the transit on the tripod head.

3 Focus the lens, as you would a telescope, on a tape measure or measuring stick. Read the height at the cross hairs.

CUTTING TOOLS

There are several different tools used to cut masonry. The most common are angle grinders and circular saws fitted with diamond-embedded masonry blades. If you plan to cut more than just a few stones, buying these blades is more economical than renting because rental stores charge a lot of money for wear and tear on rented blades. The same tool you use to cut masonry can be fitted with a metal-cutting abrasive blade to cut rebar, as well as any other type of steel or metal, such as angle iron or lath.

There are two main types of circular saws: a gas-powered saw with a 14-in. blade, and an electric saw with a 7-in. blade. I prefer the gas-powered saw: You don't have to worry about the cord getting in the way, the saw will cut just about anything, and the blade size allows you to cut large stones or concrete. Gas-powered saws are expensive, so consider renting one if you don't anticipate making too many cuts. Electric circular saws are smaller, easier to handle, and less expensive to buy or rent. Both types of saws create a considerable amount of dust. Most gas-powered saws have a fitting that allows you to connect to a water hose to reduce dust.

Angle grinders are much less expensive, but they don't have the power of larger saws. A grinder fitted with a 4-in. masonry blade will come in handy if you are cutting brick or manufactured stone or making small cuts in natural stone or block. Larger angle grinders can be fitted with an 8-in. blade.

Masonry cutting blades have tiny diamonds embedded in the cutting surface (new blade on top). Rental companies typically charge by how many microns have been worn away.

A gas-powered circular saw (also called a cut-off saw) is handy for bigger jobs but is not practical for lighter applications.

Metal-cutting blades are typically made of fiberglass, upon which an abrasive material, typically a hard mineral, is bonded.

Hand-held circular saws can be fitted with masonry blades, making them a relatively inexpensive stone-cutting option.

A wet saw, typically used for cutting tile, is a good tool for cutting brick and manufactured stone. Wet saws can be rented from your local home improvement store. Elevating them on a table makes the cutting process a little more comfortable, and most models have a convenient slide table that helps make precise cuts. Wet saws also circulate water over the blade, which eliminates much of the dust.

A brick splitter is a medium-size tool that will split brick and soft stone, such as sandstone. Consider renting one if you have a lot of cutting to do, especially if you need to keep the dust to a minimum.

Small right-angle grinders can also be fitted with masonry blades. However, because of the shallow depth of cut, their usefulness is limited.

Wet saws make precise cuts and help keep dust to a minimum.

Brick splitters require more effort than using a saw to cut brick, but there is no dust or noise to contend with.

TIPS FOR CUTTING MASONRY

There are several ways to make cutting masonry easier and faster. Whatever cutting you do, wear hearing, vision, and respiratory protection. Wearing appropriate footwear and gloves can go a long way to making cutting safer. Likewise, make sure all safety features are in working order.

Brace behind the work. To prevent a brick or stone from moving dangerously during a cut, use two concrete blocks to stabilize it.

Gang-cut whenever possible. By aligning one end, these stones can be cut to the same length in a single pass.

Keep both hands on the tool. Find a way to brace the work with something other than your hand. If a tool jumps, you will be more likely to control it with two hands than one.

Score around the perimeter if the blade isn't deep enough to go all the way through. Then tap the waste portion with a hammer to finish the cut.

DIGGING TOOLS

Much of the preparation for masonry projects happens below grade, so plan on doing plenty of digging for almost any masonry project you take on.

Hand tools for digging

The simplest and most-used excavation tool is a shovel. A common round-point shovel is good for getting the soil out of a hole in the ground. A square shovel is not quite as effective for simple digging, but I prefer it for most other tasks, such as mixing mortar, cleaning up footing excavations, smoothing concrete, or even doubling as a dustpan.

Other digging-related hand tools include rakes, mattocks, and axes. A mattock makes short work of loosening the soil when digging a footing. You can also use a mattock to pry heavy stones (saving your back and your fingers!). An ax will help you cut roots you encounter while digging.

Excavators

Heavy equipment is generally not needed for small projects. On larger jobs, backhoes and excavators make life much easier when digging a footing, moving material, or cleaning up construction debris.

To operate small earth-moving equipment, you generally don't need a license or operator's certificate. That said, if you have never been around these machines and are unfamiliar with the way they work, it is a good idea to invest in training for both safety and efficiency. Most manufacturers have training materials available through their respective websites. Some local rental stores will allow you to practice on a machine for half an hour for free. If you decide to rent equipment, be very careful—and read the manufacturer's instructions and safety manuals before you begin.

If you plan to use machinery on city streets, check the local municipality for restrictions. For instance, in some places, excavators must be equipped with rubber tracks to operate on street surfaces.

Square shovels get a lot of use on a masonry project. Here, a square shovel cleans up a trench floor and sidewall.

Small excavators can be rented by the day and don't require a special license to operate.

In addition to a square shovel, an ax, a mattlock, and a round shovel will get you through most small excavations.

Rubber tracks make this excavator suitable to drive on a street.

MOVING EQUIPMENT

Masonry supplies are heavy. A good plan to move materials around will enable you to get much more work done. On the jobsite, where using heavy equipment is often not practical, a simple wheelbarrow is the primary workhorse. A wheelbarrow will not only help you move a surprising amount of material but also provide a good place to mix both wet and dry mortar and concrete. It's worth paying extra for a good wheelbarrow—one made of thick-gauge steel with a substantial tire. Wheelbarrows used for masonry work must endure incredible abuse, and a cheap model simply won't last.

Common 5-gal. buckets help move small quantities of material from one place to another, such as from the ground to scaffolding. I use them for carrying mortar to places that will not accommodate a wheelbarrow and for dumping debris in the back of my truck when cleaning up.

In addition to wheelbarrows, wheeled carts help with moving large stones. These carts are made of steel and have large tires. Manual lifts with wheels come in handy when you are lifting heavy objects, such as heavy stones or mantels for a fireplace. Both can be rented at your local equipment store.

A small track loader (also known as a skid-steer loader) can easily do the work of three or four laborers and is easy to operate. On a small jobsite, a small track loader is especially useful to move large stones or to collect and clean up debris, including big chunks of concrete. The loader we use can be fitted with a bucket for lifting loose fill or with forks for moving pallets.

When moving a heavily loaded wheelbarrow, have one person push and another person steady the load.

A small, light-duty dump truck with a 12-ft. bed is another invaluable piece of equipment. It is small enough to fit in tight spots and can be used in muddy construction sites. Some rental stores and home improvement stores offer these for rent. If you do a lot of projects that require you to dump stone or sand, I highly recommend one. If yours is a one-time project, consider renting. For small projects, such as small block walls, manufactured stone fireplaces, or brick steps, a pickup truck will be fine, even if you have to make several trips to move materials. » » »

Five-gallon buckets are useful on almost any job for everything from mixing mortar to disposing of debris.

Skid-steer loaders can make quick work of moving piles of soil and are relatively easy to operate.

MOVING EQUIPMENT (CONTINUED)

To state the obvious, a full-size pickup truck will hold a lot more than a small truck. Full-size trucks transport wheelbarrows, buckets, scaffolding, and small pallets. My crews use 3/4-ton trucks, because they are heavy duty enough to transport 1-ton pallets of stone, cubes of block, and pallets of mortar.

A larger truck will also pull a larger trailer. Whatever size truck you currently have, an appropriately sized trailer can be a smart investment. A flatbed trailer is useful for hauling anything on pallets, including bricks, stone, block, and mortar. A dump trailer works for hauling loose debris or sand. In addition, the cost of registering a trailer is usually much less than a vehicle. If you are doing just one job, consider renting a trailer.

Front-end loaders come in a variety of sizes and can be fitted with a fork for moving pallets or a bucket for moving soil.

Small dump trucks can haul a considerable amount of building materials and have the added advantage of maneuverability.

Full-size pickups are the smallest size truck the author considers for regular masonry work.

Trailers can haul as much as a truck and they are less expensive to maintain.

DEMOLITION TOOLS AND EQUIPMENT

Demolition tools are necessary if you are tearing down masonry walls. The most common tool is a rotary hammer or small jackhammer. Demolition hammers have chisels in various sizes and are great for tearing out old masonry or concrete. If you don't use one of these often, it's a lot cheaper to rent one from your local equipment rental store.

Pry bars don't see a lot of use on masonry projects, but they come in handy for removing old stone or brickwork, especially veneers, after you loosen them with a heavy sledgehammer. Flat hoes will help remove any type of debris from a slab if you're prepping it for flagstone pavers. Mattocks can also be used for demolition work.

Jackhammers can be fitted with a variety of tips, are easy to rent, and don't require an advanced degree to operate.

Claw hammers and pry bars come in handy for demolishing wood structures and also can be used to remove stone or brick veneer.

Mattocks are useful for digging, demolition, and moving large stones.

MORTAR MIXERS

Mixing mortar by hand can be very tiring, which doesn't leave you much energy to do the task at hand. If your project is large, one that requires making more than a couple batches of mortar per day, for example, consider purchasing or renting a mixer.

I prefer a gas-powered mortar mixer with paddles when mixing mortar for stone, brick, block, or stucco because it mixes the ingredients thoroughly. Use a concrete mixer when you are using gravel, such as concrete for a footing or slab. Whichever you use, remember to frequently clean the mixer inside and out with water. Both types of mixer are usually available at equipment rental stores.

➔ **For mortar mix recipes, see p. 61.**

Motorized mortar mixers have paddles that turn the mix inside a drum; they are not designed for mixing concrete.

Electric concrete mixers have a drum that rotates on a stand and can make short work of mixing concrete.

USING A MORTAR MIXER

To make mortar, turn on the paddles first and add half of the amount of sand. Add mortar next, and then add the other half of sand. Run the paddles a few minutes ❶. When the dry ingredients look evenly mixed, add the desired amount of water ❷. Mix the batch for a few more minutes. Make sure all the lumps are crushed and that it looks like an even mixture. If you accidentally add too much water, add more of the dry mix, in the same ratio as before.

After the mortar is thoroughly mixed, disengage the gears and turn the barrel over with the handle so that the cement falls into the wheelbarrow ❸. Turn on the gears so that the paddles push all the mortar out of the barrel. After all the mortar is out, thoroughly rinse the mixer with clean water and turn it off.

SOIL COMPACTORS

If you need a soil compactor for only one job, I suggest renting a vibrator plate compactor. New compactors are expensive and are only a good investment if you use them often. A mechanical compactor isn't practical if you have a very small area to pack, such as a footing for a small wall or a column footing. For small areas, purchase a tamper. Another thing to consider is that a plate compactor will not fit into tight spaces.

Plate compactors and soil rammers, such as the three types shown here, are readily available at rental supply stores.

BUILDING UP A SOLID BASE

Base compaction is critical when installing brick pavers or a mortared patio and is the first step after excavation. Use a packer to make sure all layers of the substrate are compact, beginning with the soil ❶. Spread a 3-in. to 4-in. layer of crushed gravel with a rake and compact it with a plate compactor ❷ or tamper. Spray a little water on the gravel to help with compaction. You will know it has been packed sufficiently when your footsteps do not leave indentations. Add more stone in 3-in. layers if needed to bring the level up to the desired height ❸. After the crushed gravel is packed, spread a 1-in. layer of quarry dust evenly over the surface to bed the pavers. Use a long level to determine elevations ❹.

SCAFFOLDING

If you do any amount of work above chest height, get some scaffolding to elevate you to a comfortable working height. There are three main types of scaffolding systems used in masonry. Ladder scaffolding allows you to lift scaffolding boards as you work. This scaffolding gives you more levels when maneuvering heavy objects to the top, such as large stones or heavy containers of cement, and is commonly used in stonework.

"Walk-through" scaffolding allows you to set up scaffolding boards only at the top of each scaffold set but gives you more room to walk underneath. Stucco and brick masons generally use this type of scaffolding.

Baker-style scaffolding is similar to ladder scaffolding but is more adjustable and compact and, because it's stackable, allows you to work in tight spaces.

I recommend using scaffolding boards that are made for scaffolding, such as aluminum "clip-on" walk boards. They are much safer than lumber. Consider renting scaffolding from your local equipment rental store if you don't need it for long. Screw jacks are used for leveling scaffolding on uneven ground, and safety railings are important if you are more than one set above the ground; these items are also available to rent.

Ladders are convenient if you are doing very small jobs, such as touching up grout or plugging small holes, but I don't recommend them if you have to carry heavy objects to high places.

Scaffolding assemblies can provide exterior access almost anywhere on a structure. They are relatively safe, provided that all of the bracing and safety rails are installed.

ASSEMBLING SCAFFOLDING

Scaffolding assembly is not difficult, but it's important to check at every step to make sure the connections are fully seated and that all the braces are in place. Each set consists of two braces and two frames (or ends). The braces hold the frame together and are attached with drop locks ❶ or snap locks ❷. Each frame has four coupling pins (one on each corner), upon which other frames may be stacked ❸. Aluminum walk boards are specifically designed to fit inside the scaffold frame ❹. Other scaffolding accessories include screw jacks with base plates and safety rails. Side brackets are available if you need to extend beyond the frame.

1

2

3

4

CLEANUP TOOLS

Most cleanup tools serve double-duty on a masonry project, and it is important to clean up between project stages, not just at the end.

A small whisk broom should be in every mason's tool bag. Use it to clean small spaces and to sweep stone or brick joints after you rake them with a tool.

A push broom is necessary when cleaning up the work site or prepping a slab for flagstone. It's also good for sweeping sand into joints between pavers. Use a blower to remove dust and dirt particles from your work area before installation, and to clean a worksite after the project is complete.

Push brooms and other cleaning tools serve multiple functions on a masonry job. Brooms, for example, are used for spreading sand between pavers and for general cleanup.

SAFETY EQUIPMENT

Safety equipment is essential when you are working around cement dust, rock shrapnel, and saws. Safety training is becoming commonplace in most residential construction settings and mandatory in commercial construction settings. It takes only one small accident to see why, and you don't want to learn the hard way.

Safety equipment needed for each project is similar. Whether you are sawing, cutting, hammering, or working with cement, there is always the risk of sharp flying debris, exposure to cement dust, dropping large objects on your feet, or handling sharp objects. Basic safety equipment includes gloves, hard hats, dust masks, steel-toe boots, and safety glasses. These safety items will help prevent accidents.

Gloves, boots, and hats

Cement is especially rough on hands. Gloves will keep the cement off your skin and prevent the sharp edges of brick and stone from cutting you. Use gloves whenever handling brick, block, or stone to protect your

Latex-grip cotton gloves don't inhibit dexterity, and they protect your hands from small cuts and abrasions that are inevitable when working with stone.

hands from abrasions and also from drying out. I prefer latex-grip gloves. They are lined with cotton and polyester and have rubber coating on the palms and fingers to prevent cuts and punctures. These gloves protect your hands and help you grip objects, and the insulated version will keep your hands warm in cold weather. I like to be able to lace up my boots without having to take my gloves off, and the latex-grip glove is the best for that.

Another great glove is the rigger glove. It is thick and durable and offers the same protection as latex-grip gloves, but it is a little bulkier.

No matter what project you are doing, there is always the possibility of dropping something on your feet. Wear steel-toe boots if possible. Tall rubber boots come in handy when you are pouring concrete footings or slabs. They also are made with steel-toe protection and are great when working in muddy conditions.

Hard hats are needed when working in an area where there is the potential for falling debris. If you are doing a lot of rock hammering or chiseling, wear a hard hat to prevent injuries from shrapnel. If you are working beneath other construction activities, be safe and wear a hard hat.

Protecting your eyes, ears, nose, and throat

Always protect your eyes when using a hammer or saw to cut rocks, brick, or block. Safety glasses protect your eyes from flying debris and dust particles. They are sold with clear or shaded lenses and even prescription lenses. You can also wear safety goggles over sunglasses or prescription glasses.

Use a dust mask or respirator when cutting stone or mixing cement to help protect your lungs from dust. Ear protection is smart when working around loud heavy equipment or power equipment, such as saws and blowers. You can buy earplug- or earmuff-style protectors. Choose whichever is more comfortable.

Rigger gloves are good for heavy tasks like moving stone, and they offer a surprising degree of dexterity.

Safety glasses can have clear or shaded lenses; wear them on masonry jobs at all times.

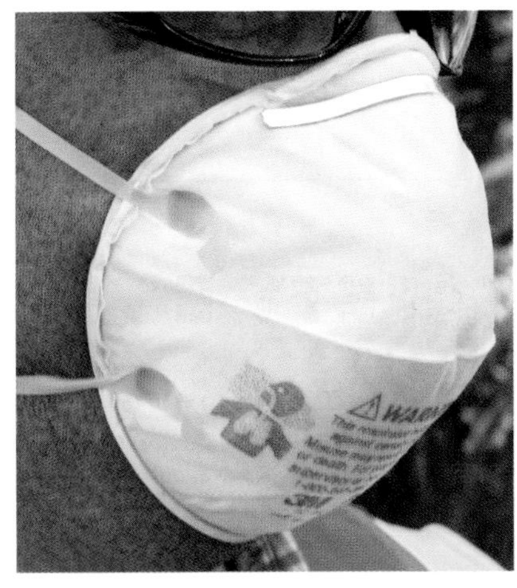

Silica-dust contamination is a real health risk. While a dust mask can be uncomfortable to wear, it's a small price to pay for healthy lungs.

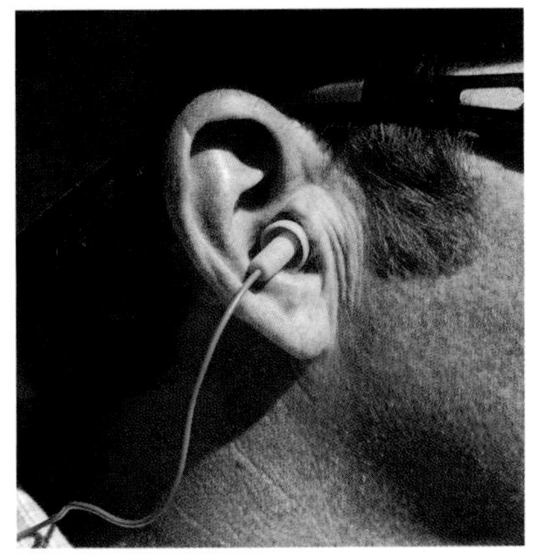

Ear protection should be close at hand. While it's not essential to wear hearing protection all the time, keep earplugs or earmuffs on you so they are immediately handy when the job gets loud.

MATERIALS AND METHODS

FOR EVERY MASONRY PROJECT, you will need several types of materials. The most obvious is what you end up seeing, called the finishing material. It can be natural stone, brick, block, pavers, or any number of other masonry products. You'll also need various cements for mixing mortar, grout, and stucco, as well as concrete and rebar for footings, and cleaners and sealers for the finished product. While it is easy to feel overwhelmed by the variety, for every project just a few tips will clarify most of the uncertainty concerning materials. This section will help you through the decision-making process and offer detailed information about each product. Whatever you end up choosing, reduce frustration by having all materials on hand before you begin work.

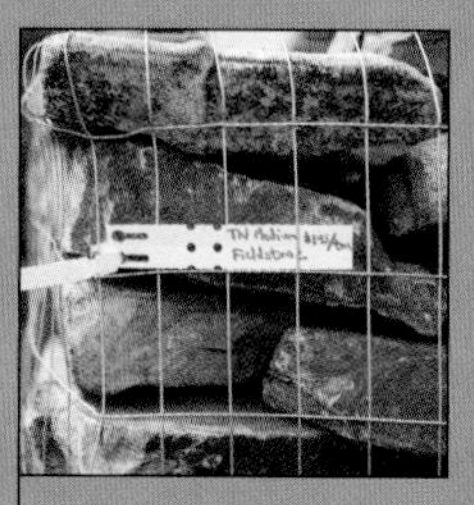

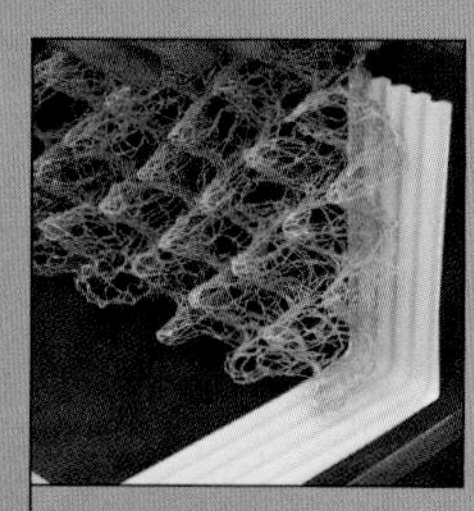

NATURAL STONE

Finishing materials include natural stone, such as flagstone and fieldstone, as well as man-made materials, such as brick, block, and cultured stone. Whatever material you choose is largely a personal preference, but each material has certain advantages and disadvantages.

Natural stone is just that—stone formed by natural forces of chemistry and pressure. There are benefits to using natural stone beyond its obvious elegant appearance. For example, you can dress the stones into different shapes with hammers and chisels. For curved walls, arches, fire pits and custom work, the ability to shape stone is crucial. In addition, natural stone has more thermal mass than manufactured stone or brick, which means it is a better insulator and can save energy costs.

The natural stone at your local stoneyard will typically have been obtained from the region where you live. Some stoneyards, however, sell imported stone alongside locally quarried stone. Local stone can be less expensive because the shipping costs are lower. Some regions, however, do not produce stone that is suitable for the projects in this book. If you feel natural stone is too expensive, consider manufactured stone.

See "Manufactured Stone," p. 56.

Natural stone includes quarried or collected stone. It's sold on pallets by the ton.

Every stoneyard will have a slightly different selection of stone to choose from. Take time to visit several stoneyards to compare prices and quality.

Fieldstone is sometimes sold on pallets bound by wire. Before purchasing, look closely at each pallet to be sure the stone is of good quality and has the characteristics you want.

Discount stone can save you a considerable amount of money, but it is often discounted for a reason. If the yard is poorly labeled (as many yards are), ask the attendant to take you on a tour. A simple tour should show you all of the available options, including discount stone, bulk stone, and palletized stone, all with different prices.

FOUR TIPS FOR VISITING A STONEYARD

Visiting a stoneyard can be a lot of fun, but buying the best stone for your project is not always easy. Here are four tips to help ensure that you get what you want and need.

1 Orient yourself. Stone is usually organized in sections, such as flagstone, boulders, fieldstone, specialty stones, bulk stone, and landscaping stones. Ask the yard attendant to help you find the section that is right for your project and limit your search to that area.

2 Look for quality, not just beauty. Stone that looks attractive in a pallet might not be the best for your project. For patios, look for stones with good edges, flat tops, and uniform thicknesses. If you are building a retaining wall, look for large, flat fieldstones. For wall veneer, smaller stones with even and consistent edges will make installation go smoothly.

3 Neatness counts. Make sure the stone is neatly stacked in pallets. Stone should be handled very few times before it arrives in the pallet. Look for stone that doesn't have chips or scratches. You can generally look at the sides of a pallet and tell what shapes are there. If the pallets are wrapped in plastic, ask the attendant to remove it so you can see the pieces more clearly.

4 Shop for bargains. Discount stone will be located in bins and is often "pick your own." However, often the bins are not marked. Never pay top price for stone out of a bin.

Going to a stoneyard to pick out materials for a project can be overwhelming because there are so many choices, but it can also be a lot of fun. In addition to standard options for walls, patios, and walkways, there will be some products that will surprise and intrigue you. It's a great place to wander as you think through your project. You can also pick out a grout color for mortared stonework, such as walls and patios.

If you are building a stone wall, your best choices are usually building stone or fieldstone. If you're laying a stone patio, flagstone or flat, thin fieldstones are excellent choices. If you're interested in saving money, ask if there are discounted products. Bulk stone is sometimes available at a lower price. If you're trying to match existing stonework or grout, bring samples so the supplier can help match them. Consider all possible options for your project before you settle on one. >> >> >>

NATURAL STONE (CONTINUED)

Stones for patios and walkways

The best type of stone to use for patios and walkways is flagstone, or flat fieldstone. Because they are flat, flagstones are easier to walk on and install. Flagstone is a sedimentary rock that can be split fairly easily into more flat, individual stones. By the time flagstone gets to the supplier and is ready to sell, it is usually split into workable pieces and stacked on a pallet. Flagstone is sold in irregular shapes and in dimensional cut sizes. Cut flagstones are squares and rectangles excellent for laying in a pattern, similar to tile. Irregular flagstone consists of varying shapes and sizes that fit together in a jigsaw pattern. Both can be used for interior and exterior patios (horizontal surfaces) and as capstones for columns and walls.

Stones for walls

When you go to a stoneyard, you'll find plenty of options for building a stone wall. You might find expensive stone, cheap stone, river stone, dressed fieldstone, thick stone, veneer stone, ledge stone, used stone, and more. All are possibilities. Good wall stones are usually thick, chunky, and close to square, with flat sides that make them easier to stack and fit together. Stones that are rectangular and blocky are the best for a wall, but they might also be the most expensive since they are in high demand. If you can find square granite for a good price, use it, but it is much more difficult to shape granite than other stone, such as sandstone. A good, hard, weathered sandstone, such as Tennessee fieldstone, is a good choice for walls because it is available in large, flat pieces and relatively easy to shape with a hammer and chisel. Avoid soft stone that is brittle and crumbly; these stones will not last very long.

Most pallets of stone, as well as bulk stone, include both large and small sizes. Some yards will separate these for you, but it may cost extra.

Characteristics of color, shape, and size can vary considerably between pallets and date of purchase. This pallet would have to be carefully mixed with the pallet behind it to blend the stark contrast in colors.

This flagstone pallet has plenty of large pieces from top to bottom and is well stacked. The stones are also of consistent thickness, which will make the project easier.

Precut stone for patios might be limited to one size or be a variety of sizes. Whatever the size, expect to do a fair amount of recutting during installation.

Irregular flagstone fits together like a puzzle and is shaped with hammers and chisels rather than cut during installation.

FITTING PATIO FLAGSTONE

The best way to fill a hole with an irregular piece of flagstone is to lay it over the opening and mark the edges that need to be cut with a pencil or the sharp edge of your hammer. Make the most difficult cut first ❶. That way, if the stone breaks in the wrong spot, you've invested less time than if you had made the most difficult cut last. With the first rough cut made, mark and shape the secondary cuts ❷. Once the stone is close to fitting, mark any necessary refinements on all the edges at once ❸. Leave a ½-in. to ¾-in. joint around all the edges, or a joint gap consistent with the rest of the patio. Trim a little bit at a time and keep testing the stone's fit until you get it right ❹. It's a good idea to have two people lift large stones.

Once the stone fits, spread mortar in the hole. Use a rubber mallet to set the stone ❺.

❶ Start with the most difficult cut, in this case an inside corner. Once you have this cut roughed out, move on to the next step.

❷ Place the stone in position and mark where the edges intersect with adjacent stones.

❸ Mark the final trim lines on all the sides so there will be a ½-in. to ¾-in. gap (joint) on all sides.

❹ Trim the edges to the marked lines. Remove small bits of stone with a light touch of the hammer.

❺ Set the stone in a bed of mortar and tap it into place with a rubber mallet.

MANUFACTURED STONE

Manufactured stone is ready to be installed as it comes from the factory, with the exception of making a few sizing cuts with a saw. Use a sharp diamond-studded blade for best results; it will give you a more precise cut. The only place on a manufactured stone you can make a precise cut with a hammer is usually in the middle.

Most manufactured stone products have warranties, which is another way of saying it will not last forever, the way natural stone will (manufactured stone is brittle). Because manufactured stone doesn't need to be dressed, it saves labor, time, and cost. You literally take the stones out of the box, butter the back of them with mortar, and affix them to the wall.

Manufactured stone is made of portland cement, lime, sand, and aggregate and is made to look like natural stone. It has become very popular because it is relatively inexpensive and is easy to work with. Manufactured stone also weighs less than regular stone and, for the most part, resembles the real thing. Today, there are many different styles to choose from, and you can buy it in almost any shape, texture, size, and color.

Manufactured stone can be used as a veneer on most vertical surfaces. Some companies even make stone shaped specifically for hearths, steps, wall caps, and horizontal surfaces. It is a popular veneer for houses, fireplaces, chimneys, and walls.

Good manufactured stone has realistic color, feels like real stone, and has an interesting variety of shapes and sizes. (I have used less expensive stone and seen the same piece several times in one small box.) Higher-quality stone also has a good warranty, such as 40 to 50 years.

➔ **See "Manufactured Stone Fireplace," p. 202.**

Manufactured stone comes in an assortment of textures and sizes and is sold by the box. Check at least one box to ensure there is a sufficient variety of stones.

Rather than picking through the yard, as with natural stone, manufactured stone is selected from samples.

The installation of manufactured stone can go very quickly, especially if there is a wide expanse of wall without windows or doors.

This section of manufactured stone fits together well, is fairly uniform, and has only a couple of repeating shapes.

BRICK

There are a few different types of brick used in masonry construction. Bricks with hollow cells are generally used for structural purposes or as veneer for walls, foundations, columns, and chimneys. The holes allow the brick to be evenly fired. Solid bricks, often called paver bricks, are used for patios and walkways. Firebricks, made of heat-tolerant material, are used for fireplaces, pizza ovens, and kilns. Firebrick is generally not suited for other applications. Bricks are sold by the pallet, strap, or individually.

For masons, the advantage brick has over stone is that bricks don't have to be shaped with hammers and chisels, except where the fit is tight. The length is generally about twice as long as the width, allowing bricks to be bonded (interlocked or woven) for structural integrity. Laying brick goes quickly once you learn the technique of setting a brick with one hand while spreading mortar with the other.

Brick is nearly maintenance-free. It doesn't need caulking or staining like wood siding, and it only needs to be painted if you want to change the color. Brick won't rot like other finish materials. Plus, it provides insulation, keeping a home cooler in the summer and warmer in the winter.

See "Brick Veneer," p. 118.

See "Brick Walkway," p. 174.

Bricks with hollow cells are used for wall construction or as wall veneer.

Solid bricks have no holes and are best used as pavers. Some rustic styles of new bricks and many old bricks have no holes in them.

BLOCK

Concrete blocks, also called concrete masonry units (CMUs), are probably the most commonly used masonry product. They are made of portland cement, sand, and aggregate. For the projects in this book, blocks are often used to provide a solid surface on which to attach a veneer of brick, stone, or stucco.

➔ **See "Block Wall with Stucco," p. 156.**

Concrete blocks are usually made with hollow centers to reduce weight and improve insulation, although cap blocks are made without a hollow core. The most common size is 8 in. by 8 in. by 16 in. The actual size is usually about 3⁄8 in. smaller to allow for mortar joints. Concrete block can be used for foundations, walls, and chimneys not in contact with combustibles. The top of the block has a greater surface area than the bottom; you spread mortar on the top. The hollow core is used for steel and concrete reinforcement.

There are also many other shapes from which to choose. U-shaped blocks are for bond beams (bond-beam blocks are designed to hold a horizontal layer of concrete to increase wall strength). Blocks with grooved ends are for control joints, or intentional vertical wall cracks that allow for longitudinal movement without affecting any other parts of the structure. Solid blocks can be used for caps; flue-liner blocks are rated to withstand contact with combustibles inside chimneys; and square blocks can be used to build columns. A wide variety of decorative blocks also exist.

Concrete retaining-wall blocks are specifically designed to stack in courses (a layer of masonry units running horizontally in a wall) on a compacted footing. These blocks are available in various shapes, sizes, and colors. They are designed to lock together when stacked in horizontal courses, with each course set back from the previous course. Check your local masonry supply center or home improvement store for availability.

Concrete pavers are generally used for patios, walkways, and driveways. They are available in various sizes, shapes, and colors and are sold by the pallet. Bed preparation for concrete pavers is the same as for a brick walkway. » » »

Concrete block comes in different colors and shapes. These decorative blocks have a textured side and can be finished with a color-matching grout.

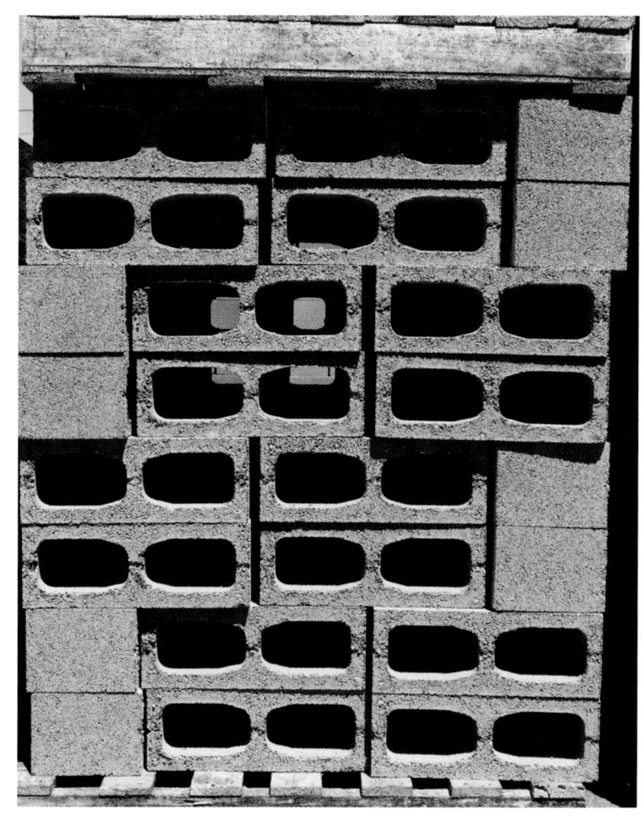

The common concrete block comes in a variety of sizes and can be used as a structural element or a finishing product. It has hollow centers for steel and concrete reinforcement, as well as weight reduction.

Retaining-wall blocks have interlocking shapes with a built-in batter. They also have a textured face to enhance their appearance.

Flue-liner blocks come in various sizes and are rated for direct contact with combustible gases.

BLOCK (CONTINUED)

CUTTING CONCRETE BLOCK

You can cut block with a brick hammer or a saw, but you'll get a more precise cut with a blade. To cut with a hammer, just tap the block in a straight line with the sharp edge of the brick hammer ❶. Keep tapping across the cut line gently until the block breaks.

To cut with a grinder, or a saw with a masonry-cutting blade, first use a pencil to mark a straight line on the block. Cut one side; then flip the block over and cut the other side ❷. If the blade doesn't reach through the block, score around the perimeter and tap it with a hammer to break off the waste piece ❸.

1

2

3

MORTAR AND CONCRETE

Mortar is a crucial ingredient in most masonry construction. It is generally used to describe the wet, sticky paste that binds together blocks, bricks, and stones when building walls, columns, chimneys, fireplaces, and patios. Mortar is typically made from cement or lime, sand, and water. You can mix mortar on the job (see mortar recipes below) or buy bags of premixed mortar ingredients at your local home supply store.

Concrete is a composite construction material made of aggregate (gravel), cement, and water. It has a high compression strength, which makes it great for footings, patio slabs, and filling blocks. Concrete is sold in bags and in bulk. For small projects, such as a stone column, I recommend buying bags and mixing the contents with water in a wheelbarrow or a concrete mixer. For larger projects, such as footings for a retaining wall or slabs that are 50 sq. ft. or more, have material delivered.

You can make stucco mortar by mixing cement, sand, coloring, and water to a pasty consistency. It also comes premixed in buckets in multiple colors. Mortar for brick and block is similar, made by mixing cement, sand, and water. To get different colors, add dye to the mix or buy premixed, colored mortar. Mortar for stone is similar. It is made of a combination of portland cement, sand, and water. >> >> >>

MORTAR RECIPES

Cement, sand, and water make mortar. Recipes will vary based on the type of mortar and the amount of each ingredient you use. Use the following recipes for the projects featured in this book. Different amounts of water are required depending on the job and weather.

- For brick or block, use 1 bag of Type S masonry cement, 18 shovels of sand, and water.
- For dry-stack stonework, use ½ bag of portland cement, ½ bag Type S masonry cement, 12 shovels of sand, and water.
- For jointed-style stonework and flagstone, mix ½ bag portland cement, 12 shovels of sand, and water.
- For thin-veneer stonework, mix 1 bag Type S masonry cement, 8 shovels of sand, and water.
- For stucco, mix ½ bag portland cement, 12 shovels of sand, and water.

Masonry cement and portland cement are the key ingredients of mortar.

EASY-OPEN BAG

To reduce dust and muscle strain, open a bag of cement by first placing the unopened bag where you want to mix the contents. Then use a shovel to open one end. Empty the bag by elevating the unopened bag end as smoothly as possible. To reduce dust, have a helper lightly spray the contents with a hose as it exits the bag.

HAND-MIXING MORTAR

On some projects, using a mortar mixer is not practical. The terrain may be too steep or too rough to stage a mixer near the work. Maybe you don't have a vehicle that will tow a mixer. Or perhaps the job is too small to justify the expense. Whatever your reason, sometimes hand-mixing mortar is the easiest way to go. There are two good methods for hand-mixing mortar: in a wheelbarrow, and on the ground on a clean sheet of plywood.

Mixing mortar in a wheelbarrow

Mixing mortar in a wheelbarrow keeps the materials contained, reduces the labor of moving materials, and is kinder to your back than mixing it on the ground.

To mix in a wheelbarrow, shovel half the sand into the wheelbarrow ❶, add the cement ❷, and then add the rest of the sand. Use a shovel (preferably with a square blade) to pull the dry mixture toward you in the wheelbarrow, chopping with the shovel while pulling ❸. When all the mix has been moved to one end of the wheelbarrow, walk to the opposite end and pull the dry mix to that side. Repeat this until the dry ingredients are thoroughly mixed. Add a couple inches of water to the free space in the wheelbarrow ❹. Then, pull the dry mix through the water using the same chopping motion with the shovel ❺.

Repeat the process of moving the mix from one end of the wheelbarrow to the other until the water is evenly mixed in. Add more water in small increments, mixing each time, until the mortar is the proper consistency. Each project is different and requires a different amount of water to achieve the appropriate consistency ❻. >> >> >>

1 Shovel in half the amount of sand. Try to make each shovel load consistent.

4 Add a few inches of water to the wheelbarrow's open side (don't add all the water at once).

TRADE SECRET

It is always easier to add a little more water to a too-dry mixture than it is to add sand and cement to an overly wet one. Each batch is unique with regard to the amount of water needed, but you will begin to get the feel of how much to add after a little practice.

2 Add the cement. Empty the bag slowly to keep dust to a minimum. Then add the rest of the sand.

3 Mix the dry ingredients by pulling the shovel in a chopping motion. Move all the contents from end to end.

5 Mix the wet and dry ingredients by pulling the sand and cement mix through the water. Avoid lifting and turning the mix.

6 Continue mixing until the mortar has the desired texture. Add water in small amounts if the mix is too dry.

HAND-MIXING MORTAR (CONTINUED)

Mixing mortar on plywood

Mixing on plywood is similar to mixing in a wheelbarrow, but it gives you a bit more room to maneuver a shovel and mix the ingredients. Also, two people can mix with shovels, which makes the process go twice as fast. Spread the sand in a pile in the center of the plywood. Form a shallow crater in the sand, add cement and spread it evenly around the crater ❶. Then shovel the sand into the crater center until all the cement, and sand is in one pile ❷. To thoroughly mix the dry ingredients, move the pile from one side of the plywood to the other, and then back again ❸. The mixing is done when the dry ingredients are a uniform gray color without significant pockets of sand or cement ❹.

To mix in the water, form a new crater and fill it about halfway with water. Use the shovel in a chopping motion to draw the dry mortar from the edges into the water ❺. Be careful not to let the water escape from the pile. After you shovel all the dry mix into the center, keep mixing the mortar by shoveling from one side of the plywood to the other, in the same manner as you mixed the dry ingredients ❻. If you need to add more water, use a sprayer and lightly mist it, just a little at a time. Keep mixing until you reach the desired consistency ❼, which is specific for each project.

1 Create a small crater with the correct amount of sand and then spread the cement evenly on top.

4 Mix the dry ingredients until the sand and cement have an even color without concentrations of either material.

7 Stop adding water when the mortar has the right consistency for the job at hand.

2 Pile the dry ingredients in the center to mix the sand and cement.

3 Move the pile from one side of the plywood to the other, lifting and turning each shovel load.

5 Form a small crater in the dry mix and add a few inches of water. Start mixing by pulling the shovel across the crater, chopping as you go.

6 Lift and mix the mortar, adding small amounts of water as needed.

Type S vs. Type N Masonry Cement

Type S masonry cement has a higher compressive strength than Type N. It is used for structural purposes, such as load-bearing concrete masonry unit (CMU) walls. It bonds to brick, stone, and block. Type N masonry cement is nonstructural, which means it will not bond as tightly to masonry. The advantage to Type N masonry cement is that if the wall cracks slightly, the mortar joints act independently of the brick. The joints will crack, but the brick won't. The joints can be regrouted.

Masonry reinforcement is essential and sometimes mandatory when building masonry structures. Movement due to changes in substrata, freeze-thaw cycles, and underground moisture can cause settling and cracking damage. Walls must be built using some type of stone support and masonry anchoring systems. Rebar is used to anchor a wall to a footing or slab. Reinforcement is needed between courses in brick and block work, too. Wall ties are used to securely attach veneers to walls. Below are some reinforcement options for your masonry project. When purchasing these materials, use stainless steel if possible.

Masonry strength is substantially increased by metal reinforcement. Here, vertical rebar combined with wire mesh adds tremendous strength.

WIRE MESH AND LATH

Wire mesh is used between block coursing and in concrete slabs and footings. A wider mesh is also available for composite walls. Some anchoring systems for block and veneer combine mesh and anchors; these are usually more expensive. Wire mesh is also used to reinforce concrete slabs to prevent cracks when the concrete shrinks. It is most important to prevent cracking in concrete slabs or patios if they are not going to be veneered. But even if you are covering the patio with brick, stone, or tile, cracks in the concrete can affect the finished surface, ultimately causing cracks in that, too. To be on the safe side, use reinforcement.

Metal lath is used to attach cultured stone or stucco to walls. A moisture barrier is installed on Tyvek® or a similar building wrap first, followed by lath and a scratch coat. Use nails or staples to attach lath to the wall. Veneer stone or brick is then attached to the scratch coat. Stucco has the same layers but instead of stone and brick, a brown coat is applied to the scratch coat, followed by a finish coat. Lath is necessary to hold all of these layers together. It is available in different thicknesses and shapes, including corner beads for making straight corners.

See "Applying the Brown Coat and Stucco," p. 172.

Wire mesh embedded in the mortar between block courses is easy to add and relatively inexpensive.

Metal lath helps secure stucco to the wall and is installed with wide-headed screws or nails.

WALL TIES AND ANCHORS

Wall ties and anchors secure new masonry walls and veneers to existing walls. Without them, water that gets between the two walls might cause the two walls to come apart over time. Wall ties can be as simple and inexpensive as single corrugated strips of metal, or they can be sold in pieces that require assembly. Ties may be nailed to existing walls or inserted into the mortar joints during construction to keep the brick or stone veneer attached.

Simple wall ties for stone or brick veneer are inserted into joint mortar between concrete blocks.

INSTALLING LATH

For metal lath installation, make sure you have a pair of metal-cutting shears or snips, nails or screws, and a hammer or screw gun. If you are working on a large wall, an air compressor and nail gun will save you a lot of time. For the project shown here, the builder installed a masonry backerboard on the fireplace, so we used 1-in.-long screws and a screw gun to attach the lath.

First, make sure the wall surface is clean. Use a brush to remove dust particles or debris. Cut the lath to fit with metal-cutting shears or snips. You can also use a grinder that's fitted with a metal-cutting blade. I usually start from the bottom and work up. Overlap lath 1-in. minimum on all sides and space the fasteners 7 in. apart, unless no sheathing is present, in which case you can space them 16 in. apart over the studs. The lath sheets should always be installed horizontally, perpendicular to the framing.

Use 1-in. screws to attach lath to masonry backerboard, 1¼-in. roofing nails to attach it to wood sheathing or through drywall into studs, 1-in. concrete nails to fasten it to masonry walls, and 1-in. self-tapping screws to attach it to metal framing.

REBAR

Rebar is used to increase the tensile strength in reinforced concrete and masonry structures. It is available in various lengths, thicknesses, and grades and is used to reinforce concrete slabs, such as patios and sidewalks; to reinforce footings for walls and columns; and to attach block walls to footings and slabs. Reinforced masonry bricks and blocks have voids made to accommodate rebar. Rebar sizes are called out by numbers based on metric or English measuring systems. In the English system, each number represents 1⁄8 in.; thus, no. 4 rebar is 4⁄8 in. thick (1⁄2 in.).

To cut or bend rebar, it is best to use a specially designed tool. A rebar cutter looks like a large bolt cutter, which is laid flat on the ground. It is designed to cut and bend most rebar sizes. To cut rebar, place it in the cutter and push the long handles until the rebar snaps in two. The greater the rebar thickness, the harder it is to cut. (Rebar can also be cut with a chop saw or grinder, if you have a metal-cutting blade.) To bend rebar, insert it between the pins on the side of the rebar cutter and push the handle until the desired bend is attained.

Rebar can be purchased in a variety of widths and lengths. It is also available pre-bent, as shown here.

Control joints Control joints provide a vertical separation between two wall sections to allow for longitudinal movement. If the wall settles, the cracks are isolated rather than spread out. If you have ever seen a vertical crack in a brick or stucco wall, you have seen firsthand why control joints are important. Control joints, filled with different types of flexible caulk, are established in both vertical walls and concrete slabs, such as sidewalks and driveways.

Cutting rebar on site is simple with a cutter that can be found in most rental supply stores.

Pins on the side of the rebar cutter can be used to bend rebar to a desired shape. Do not plan on using a rebar cutter-bender on rebar larger than 5⁄8 in. in diameter.

BARRIERS, FILTERS, AND MEMBRANES

There are several synthetic materials that enhance masonry installation performance. Moisture barriers are used behind brick, stone, and stucco veneers to stop moisture from getting into the structure. Whenever applying veneer to a building, attach a moisture barrier to the wall first. The most commonly used barrier is asphalt-based paper (or roofing felt). Use a minimum 30-weight (30-W) felt on the exterior and 15-W felt for interior applications. Felt is sold in rolls and should be overlapped and attached with nails or staples.

Drainage mats and cavity wall systems are used behind stone veneers to ensure that any water that does get behind the veneer can drain away. Attached to the wall with staples or nails, they channel water through the wall cavity to the bottom of the wall where it can exit through weep holes. Drainage mats specifically designed for stone are made with a filter-like mesh that prevents the mortar squeeze-out from accumulating at the base of the wall and clogging up the cavity. These products are sold in rolls and can be found at any masonry supply store.

Antifracture membranes are sometimes used under stonework to prevent potential cracks from penetrating from the substrate to the stone. Since most stonework is laid on a concrete slab, that is where we apply the membrane, which is also called underlayment or crack-isolation membrane. Antifracture membrane is sold in rolls for exterior and interior applications. It is paired with a primer, which must be applied to the slab before the underlayment is laid. Make sure the slab is clean before applying the primer. Antifracture membranes not only isolate cracks but also provide a waterproof layer between the stone and the slab. By reducing the amount of moisture under the stone, the slab will likely maintain its crack-free integrity longer.

Weep holes direct water from one side of a masonry structure to the other. They come in various shapes and sizes, including 4-in.-long plastic tubes with cotton wicks, flexible PVC louvered weep holes, and a honeycomb design that restricts the ingress of insects and other debris. Different options work with various projects; consult your supplier for advice. I prefer honeycomb-style weeps for brick and round plastic weeps with wicks for stone.

Water that gets behind masonry **(and there will be some) needs a place to escape. Here, a combination of moisture barrier, mortar net, and weep holes directs moisture away from the masonry structure.**

Antifracture membranes **help prevent patios and walkways from cracking and also offer some degree of moisture control.**

Drainage mats allow water **to flow out freely but restrict the passage of soil and aggregate.**

Weeps positioned **in mortar joints allow water to drain.**

PLANNING AND PREPARATION

THE MOST IMPORTANT WORK, THE work that will make your project run smoother and end with a better result, happens before you start building. It's not always possible to iron out every detail before the project begins, but whatever you can take care of beforehand is one less thing to deal with later on.

This chapter covers the basics of preparation and can be used as a reference guide as you plan your projects. That said, each project is different and each region of the country presents its own challenges, whether it be strict code laws or limited resources. But with the planning skills presented here, you'll be aware of how to lay the groundwork for most residential masonry projects.

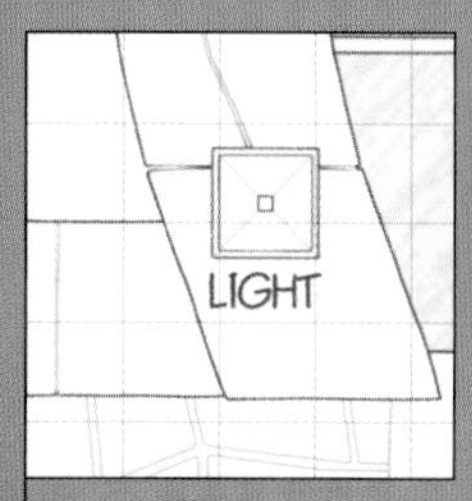

DOUBLE A
AA
BODY BUILDERS
QUIKRETE
PORTLAND CEMENT TYPE I/II
ROANOKECEMENT

DRAWING A SITE PLAN

Planning on paper can include everything from designing and drawing a site plan to scheduling equipment rentals. This chapter will help most if you've already read "Designing with Masonry" (pp. 2–17) and have decided on what masonry materials you'll be using and what the general scope of the project is, that is, small patio, retaining wall, walkway, and so on.

Drawing an accurate site plan is a critical step. It will help you do everything from estimate materials and design construction details to establish staging areas and schedule material deliveries. You do not have to be an artist or architect to create a good site plan. To make a site plan, all you need is a pad of graph paper, a pencil, and a measuring tape. For 1-cm. graph paper, I let each square represent 1 sq. ft., but you can adjust this ratio for smaller projects. Measure your project area and transfer the landscaping details, such as buildings, trees, and existing paths, to the pad as if you were looking down from an airplane. This is called a plan view. If you know the location of underground water, electrical, or gas lines, draw these on the plan as well.

The more detail you add to the plan, the more useful it will be. Consider drawing an elevation view and a cutaway view. An elevation view shows one side of a structure in two dimensions (no perspective required). A cutaway view shows a similar view but slices through the structure, including whatever is below grade. You may want to draw several elevation and/or cutaway views if the project is complex. Even for patios and walkways, a good cutaway drawing can help when calculating footing size and material requirements. Additionally, you can take your plan to your material supplier to help communicate what you are going to need for your project.

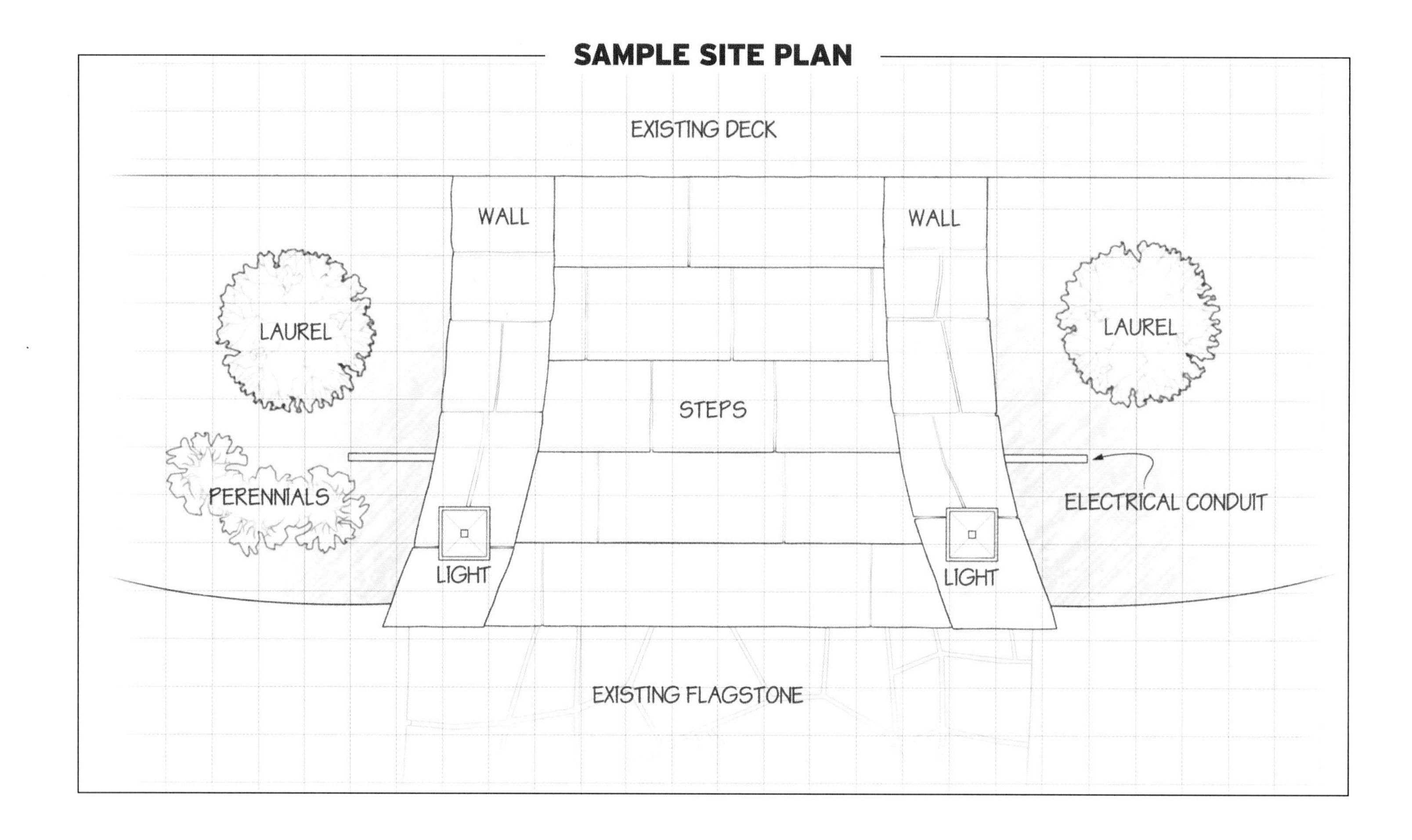

ESTIMATING COSTS

With a good site plan, you should be able to estimate your material cost just by determining how large your job will be. For example, if you can calculate a wall face's square footage, most suppliers will be able to tell you exactly how much stone you will need. This is also true for brick, cultured stone, and stucco. Measure wall caps independently if you are using a different type of material.

A lot of construction material prices fluctuate, so ask your local supplier how to get the best price. This may mean ordering ahead of time, or perhaps paying in advance to lock in a price. Most suppliers will hold the material for you for short periods for free, but they will probably need to get paid to secure the material for long periods.

Sometimes all you have to do is ask for a good deal. There is a lot of profit margin in many items. If a supplier thinks you might be shopping around, you might get quoted a better price. Some suppliers frequently liquidate material, especially in a slow economy, so it doesn't hurt to ask to see liquidation inventory. One person's trash can be another person's treasure. I recently purchased material that was in inventory for a couple of years for 50 percent of original cost.

SAMPLE ESTIMATING SHEET

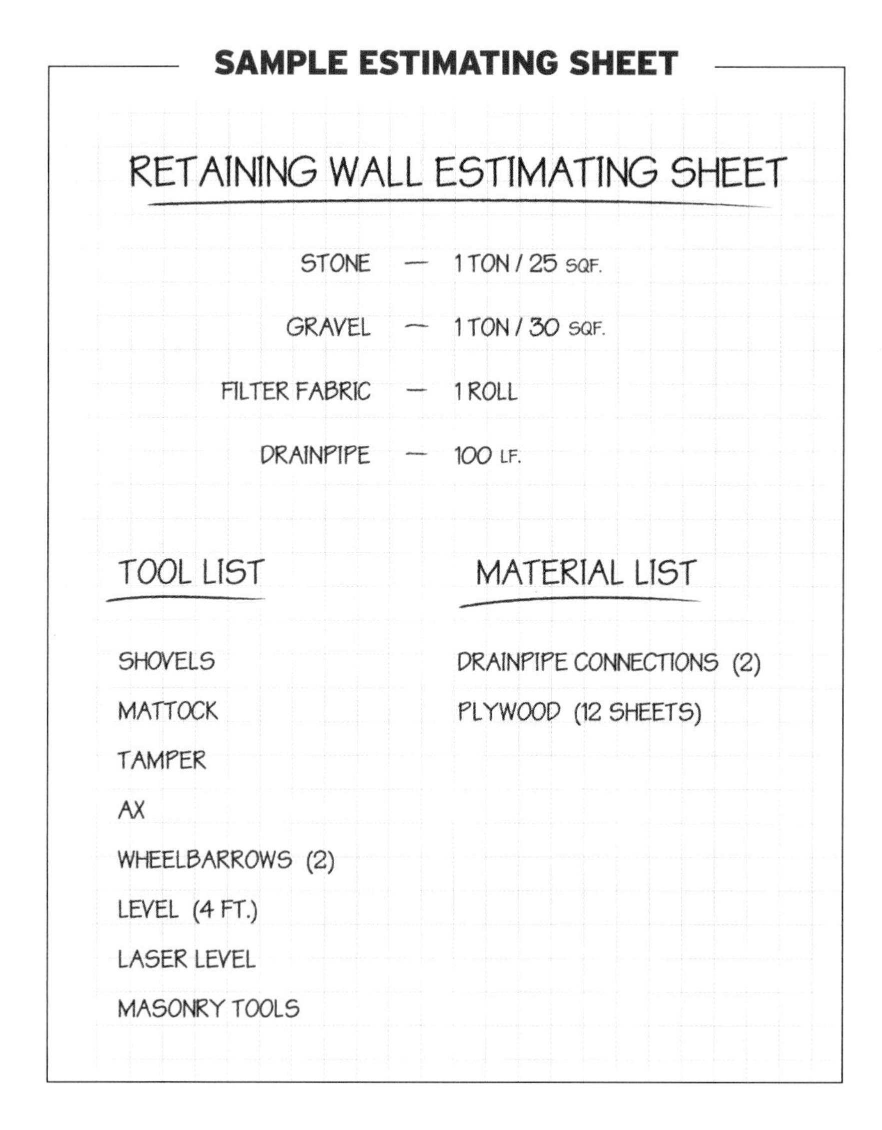

PREVENTING EFFLORESCENCE

Efflorescence is a white deposit that occurs mostly on brick and stonework when water escapes from the surface or the joints. It evaporates and leaves salt behind. It is mostly an aesthetic concern, not structural. Light efflorescence can be removed with a nylon or wire brush. Heavy efflorescence requires a pressure washer or sand blaster.

Efflorescence is more common in areas that get a lot of water and in some cases it is impossible to prevent. Applying a sealer helps, and I have also found that adding more water to the mortar alleviates the problem.

ESTIMATING TIME TO FINISH

I am often asked, "Can you finish our project by next week?" Coming up with an answer gets easier with experience, but even after doing masonry work for many years, it's still a difficult question. Small projects that take one to two days are easy to predict. Larger projects that take a week or longer are subject to a lot of variables. As a general rule, the longer a project takes, the more difficult it is to confidently predict a finish date. That said, once a project begins, you can establish a baseline for how quickly the work can get done. With larger projects, it may be helpful to keep a log or journal, to remind yourself of the progress you are making if nothing else.

TALKING TO THE PROS

There are times when it is a good idea to consult a professional. For example, retaining walls over 4 ft. tall should be designed by an engineer. You may also want advice on how to handle drainage issues, code requirements, construction methods, or other issues that arise on a masonry project. Some people are reluctant to involve professionals because of the expense, but this often ends up costing more down the line.

Bear in mind that seeking out professional advice is not an all-or-nothing deal. Often, an architect or licensed contractor will meet with you to discuss your project and waive the fee for the initial meeting. And you may find that continuing the consulting relationship is more reasonable in terms of cost than you initially thought.

Another source of information is your local building department. Some local building codes require you to submit an application, and the department can be an invaluable source for technical information. When in doubt, contact your local building inspection department for information.

In an effort to save money, you may encounter some unscrupulous tradespeople who like to fly under the radar when doing small construction projects on personal property. Whether it is an unlicensed contractor or a mason who encourages you not to apply for the proper permits, these people should be avoided. Getting involved with someone who likes to cut corners will invariably cause more headaches than the savings are worth.

Secure Help if Needed **Masonry work can be physically exhausting. Moving blocks all day, digging footings, and mixing mortar is taxing on your body. Hired help will always make the job easier and quicker to finish. If you are doing mundane manual tasks, such as the ones just listed, you will not need skilled labor. If you are laying stone or brick or applying stucco, it will benefit you to hire someone skilled in these trades. If not, you will be spending all your time as an instructor.**

For tasks such as lifting heavy stone mantels or other large objects, you will almost certainly need a helping hand. Consider asking a neighbor or a friend if the task doesn't require too much time.

PROTECTING THE SURROUNDINGS

Depending on the specifics of your project, you could have extensive preparation work to complete—or hardly any at all. No matter what the project is, however, you'll want to create a preparation checklist and address everything you can before the project begins. It is annoying and inefficient to stop mid-project to complete a task that could have been done beforehand.

Masonry projects typically involve working with significant quantities of mortar, cement, dust, gravel, and possibly flying shards, all of which can be messy and potentially damaging to adjacent plants and buildings. Protecting the surroundings prior to material delivery or staging will save a lot of frustration.

The first line of defense against damage is a roll of heavy-duty polyethylene plastic. I recommend 6-mil thickness, because it is thick enough to handle some abuse, yet thin enough to work with easily. Spread the plastic over the staging areas for construction materials, such as sand, concrete bags, and other masonry materials. If you are planning to mix mortar on the ground, spread plastic underneath a sheet of plywood and mix the mortar on the plywood. This will give you a flat surface to work the shovel against and fully protect the area where you are mixing. You can use the plastic to cover everything from nearby shrubs and garden furniture to sealing doorways and air ducts. If there's the threat of rain, use the plastic to cover your sand, stone, brick, and mortar, so they will be dry when you are ready to work. >> >> >>

Whether staging materials on a driveway or lawn, spread a single layer of 6-mil plastic under your materials to avoid scrapes and stains.

This staging area was established so the author's crew could drop off materials easily.

Laying plastic under the work zone enables you to clean up in a matter of minutes at the end of the day, leaving the job site clear of dust and debris.

PROTECTING THE SURROUNDINGS (CONTINUED)

If you are working on grass, spread a layer of plastic so you can cut stone and mix mortar on top of it and prevent the small pieces of stone and mortar from getting into your lawn. Choose a shaded area: Covering grass that is in direct sun is a good way to kill it off in just a couple of hours. If there is no shade, consider an inexpensive canvas canopy or stage in an area where there is no lawn.

Another important item to have on hand is protective tape. It's a good idea to tape areas that are adjacent to mortar, such as windows and trim. When building a fireplace with a metal insert, tape the edges of the metal so that you don't get cement on it. Blue painter's tape or Frog Tape® is designed for masking and won't leave a sticky residue.

Protect finished surfaces with plastic and sheets of plywood, cardboard, or hardboard. If you are building a fireplace in a finished living room, for example, spread 6-mil plastic on the floor (and over the furniture), and then place plywood on top of the plastic. Plywood is a great surface for stacking tools and rocks, and it provides safe storage for buckets of mortar. Make sure to sweep thoroughly before laying the plywood down because any particles that get trapped under the plywood can mar the floor's surface.

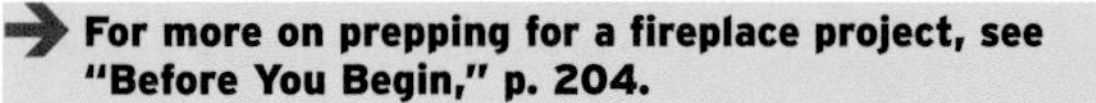
For more on prepping for a fireplace project, see "Before You Begin," p. 204.

Plastic under the sand not only protects the driveway, it also prevents the sand from absorbing excess ground moisture. On this particular day, the forecast called for overnight showers, so the author's crew covered the top as well.

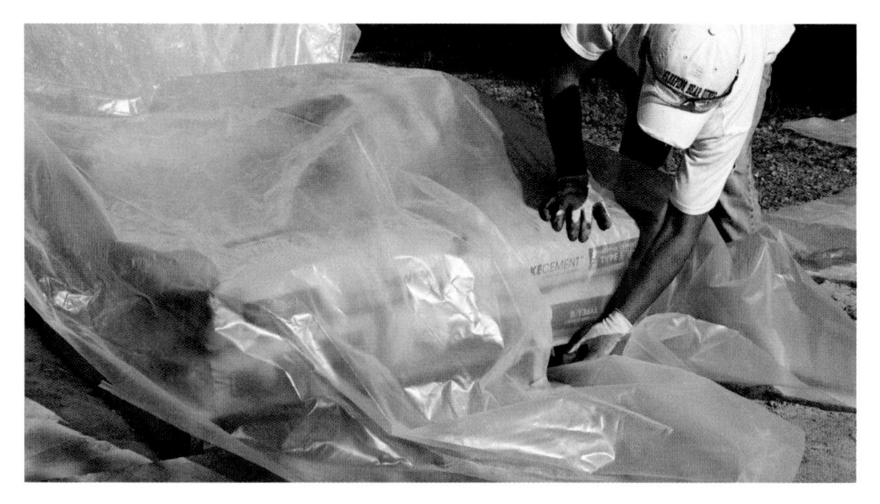

It is extremely important to cover bags of cement or concrete mix to prevent water from contaminating the bags. Even a couple days of heavy dew can ruin the contents of the bags.

If you're working indoors, it is better to err on the side of too much protection. Even if a floor is unfinished, protect it with plastic and either cardboard or plywood.

ESTABLISHING STAGING AREAS

If you are going to gather tools and materials ahead of schedule and allow yourself plenty of time to shop for good prices, you are going to need to establish a staging area (or areas).

For some projects, such as stucco and small jobs, your staging area can be very small, such as the corner of the garage. But for projects that require a lot of material, such as stonework and brickwork, you will need a large staging area or areas. These projects often require mounds of stone and sand and pallets of cement and brick. Not only do you need room for these materials, but you also need space to move them around and to mix mortar.

If you are veneering an exterior house wall, it is a good idea to remove shrubs and other small plants when possible, as well as other objects that might get in your way, especially if you are using scaffolding and heavy equipment. On one job, we temporarily removed recently installed sod because it rolled up so easily and thus wouldn't be damaged.

If you are working indoors, remove any objects that will be in your way. You want to make sure you have plenty of room to stock material, build scaffolding, and of course, do the work. If you don't have the space indoors, rent a storage container to use as a staging area. Storage containers offer secure equipment storage and help keep the mess to a minimum.

If a staging area close to the project is not possible, you can set up a remote staging area. You can always move materials closer to the site with a truck or other moving equipment.

The more organized you can be within your staging area, the more likely your whole job will be organized.

Sometimes it is better to establish a staging area for ease of drop-off and then use a wheelbarrow to haul the stone to the project area.

If you can stage materials adjacent to the work zone, so much the better. Close proximity can speed the work.

FROM PLAN TO GROUND

Before beginning work, you need to transfer what is on your site plan to the project area. You can use marking spray, wooden stakes, stringlines, or batter boards (or any combination thereof) to lay out the site. Start with a fixed object on the site plan, such as a tree or the corner of a building. Then measure to your project's perimeter, whether it is a patio, walk, or wall. Try to establish more than one point of reference to double-check the location of your layout. At each crucial point or intersection, I make a dot with marking spray. When I'm satisfied with the layout, I connect all the dots to create the project outline. Take the time to get the measurements right before continuing because site layout is critical to starting your project correctly.

If you've completed the site layout and still have lingering questions about property lines, easements, or underground lines, consider having a site survey done. This can be expensive (possibly running into thousands of dollars), but that's preferable to building in the wrong spot or, worse, rupturing a utility line.

For more challenging layout projects, such as a freestanding wall, consider finding a helper to get the job done. Two people can be more than twice as fast as one because of the time you save not having to walk back and forth across the project area (or not having to secure one end of a tape measure every time you want to pull a measurement in the grass).

If you are building steps and are having a difficult time laying them out, create a story pole. Once you figure out the number of risers and treads you need, mark each step on a long, vertical 2x4 or 2x6 board and use it as you lay block and stone or brick. Instead of measuring and marking each time you reach a new step, just move the pole as you work (see the sidebar on p. 142).

Another important layout technique is the use of batter boards. Batter boards are designed to stay in place throughout the entire project. They are used to reference excavation, edge concrete forms, and lay masonry. If you need to be accurate with your measurements over multiple construction phases, use batter boards. While setting up batter boards can take longer than pulling a simple stringline, they save time and avoid possible measuring errors as the project progresses (see the sidebar on p. 34).

For a rustic wall, a simple stringline serves as a useful guide during site layout and the first phases of construction. More complex jobs require batter boards.

When laying out the site, it's sometimes helpful to have several people to hold tape measures, stringlines, and levels. This will leave you free to look at the plans.

When first laying out a project, mark corners and intersections with dots of spray paint.

Marking spray is a valuable tool when it's time to draw the project layout on the ground.

After you've double-checked that the initial marks are correct, connect the dots to complete the layout outline.

PREPARING FOR EXCAVATION

With almost every exterior masonry project, some excavation is required. You may have to dig before pouring a slab for a flagstone patio. Or you may need to dig out a trench to pour a footing for a retaining wall. You may even have to dig down to an existing house footing to stack a stone veneer. During the preparation phase, decide how to accomplish the excavation and get the equipment or people scheduled.

The most obvious preparation for excavation is simply to secure a shovel and a pair of gloves. You can get away with digging by hand if it's a small project. It's surprising how much soil you can move in an afternoon. If you can convince friends to join in, four or five strong diggers can easily match the work of a skid-steer loader.

If the job requires greater excavation muscle, there are two choices: Rent equipment and operate it yourself, or hire someone to do the excavation for you. If you decide to rent equipment and operate it yourself, call your local utility companies to send a locator to mark any underground wiring before you begin. Check with the city to see if there are any public utilities underground. Also, check site surveys for old utility lines that may not be marked on other maps. In addition, check to see if any permits are necessary before you begin.

If you are working in or around a public area, there are other things you should consider. For example, you may need to install a temporary safety fence (around the site) and put up a "sidewalk closed" sign. This sign can be as simple as spray-painting a piece of plywood, but it's important to let the public know to stay away.

For any project that requires significant excavation, put up a silt fence to prevent erosion. A fence isn't necessary with a small column or patio, but if you are building a large retaining wall and there is rain in the forecast, install one to prevent erosion and excessive water runoff. Building codes require you to do this in most places.

Whenever you are working in a public area, post notices and use caution tape so that no one is injured on your materials, especially small children.

A small excavator can do the work of several diggers and can be rented by the day.

If you are going to operate an excavator or any other equipment, it is a good idea to have a helper. Agree on hand signals before you begin so you can make yourselves understood amid the noise.

If the project abuts a city street or sidewalk or requires you to close off a sidewalk, set up temporary safety fencing (and check city ordinances).

Hiring an excavation contractor can sometimes be the way to go. If you do hire one, remember that good communication is crucial. Undoing mistakes is very costly.

In areas with embankments, erosion-control fencing may be required to prevent rain from washing soil onto a city street.

DRY-STACK RETAINING WALL

RESIDENTIAL CONSTRUCTION CAN leave behind abrupt changes in grade elevation. Over time, banks of soil become rutted by erosion and host to scraggly vegetation. Every rainstorm washes soil over lawns, driveways, and flowerbeds. As the name implies, a retaining wall keeps the soil in place. Just as important, a retaining wall adds beauty, order, and curb appeal to the overall landscape around a home.

A dry-stack retaining wall is built without mortar. It uses gravity and a vertical pitch to hold itself, and the soil behind it, in place. In this chapter, we'll start with foundation details and then run through the steps of installing filter fabric, adding drainage, setting stones, and finishing the wall with large capstones.

PRACTICAL CONSIDERATIONS

Dry-stacking a stone wall can take as much time as building a mortared wall because each stone takes more time to fit and each void has to be filled. That said, it is easier to build and repair a dry-stack wall because there's no need to mix mortar and stage those materials and you can spread the work over several weekends. Of course, you don't want a job to drag on too long. If you don't buy all the stone at once, it may be hard to match the new stone to the old. And leaving an embankment unprotected is like neglecting an open wound; you may have erosion problems and a dirty mess on the driveway after a hard rain. Cover unfinished work with plastic or mulch to prevent erosion.

Consider the scope of the project and decide whether or not to consult a professional or hire skilled labor. If the wall you're planning is more than 4 ft. tall, it probably needs to be mortared rather than dry-stacked on a concrete footing. Also, any wall over 4 ft. should be engineered. An engineer will be able to account for frost heave, soil compaction, hydrostatic pressure, and other factors that affect the wall's structural integrity.

Estimating how long a wall will take to build is a simple calculation. A skilled mason can build about 20 sq. ft. to 30 sq. ft. a day. An efficient DIYer can expect to build 10 sq. ft. to 15 sq. ft. on a good day. So a wall that is 3 ft. tall by 10 ft. long will take an experienced waller and a helper one day to build. It will take a DIYer about three days.

For the project described in this chapter, the homeowners wanted to fix a wall that ran along the inside of their circular driveway as it dropped down from the street. The old wall was a good example of problems that arise without proper drainage and soil retention. With every rain, soil washed through the wall and onto their asphalt drive. Hydrostatic pressure dislodged stones and caused the wall to bulge in places, giving it an unorganized and unstable appearance.

The best solution was to tear out the old wall and rebuild with proper soil filtration and drainage. Because the new wall would be slightly bigger, we decided to replace all the stone rather than try to patch and match. The old stone ended up being used for the footing and for smaller walls on the same property.

RETAINING-WALL DRAINAGE

Hydrostatic pressure is the force that's exerted behind a retaining wall due to the accumulation of water. Over time, it can push the wall forward and weaken it (as shown in the drawing at left below). Providing a place for the water to escape can prevent this problem.

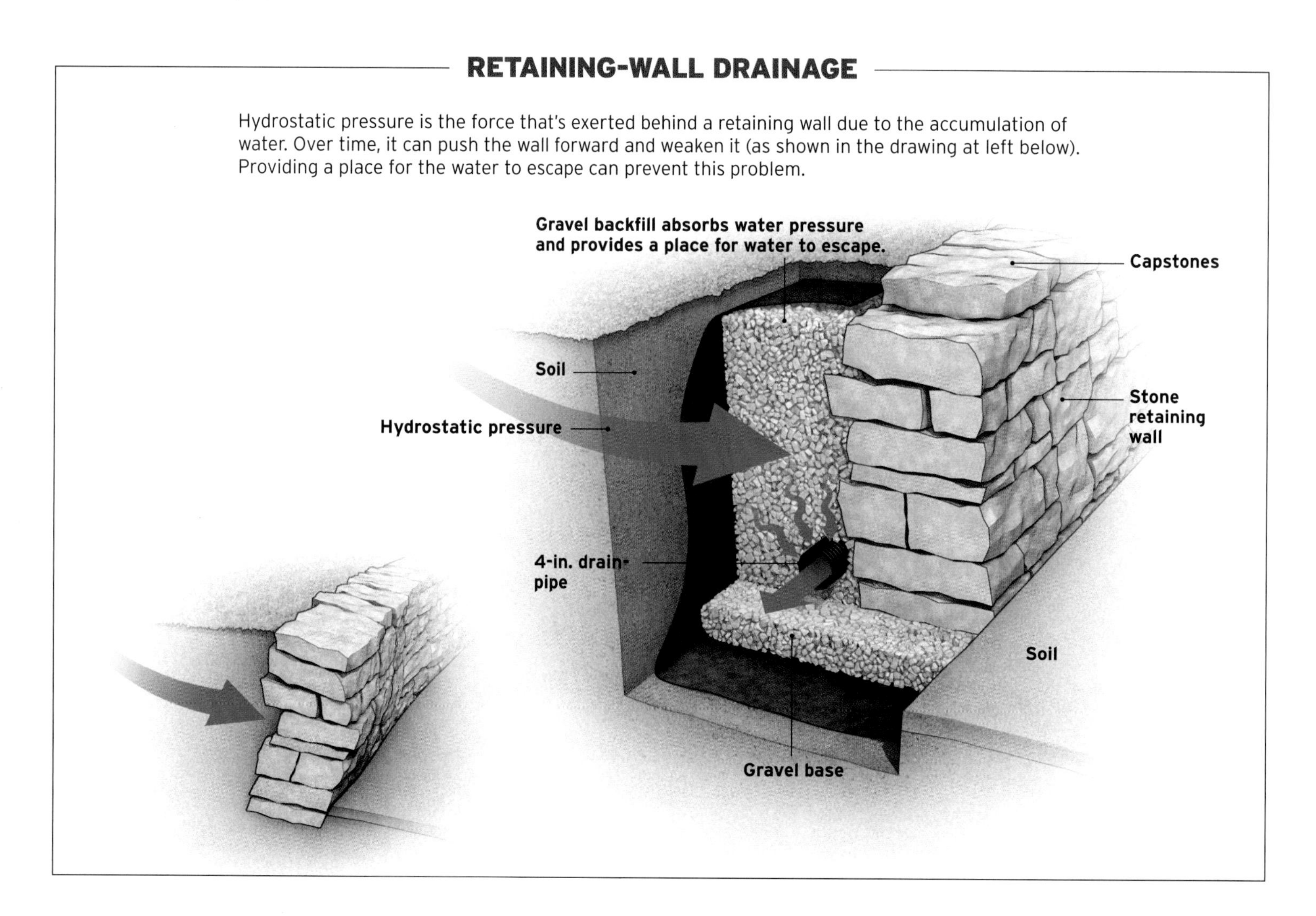

The author carefully sets **a capstone on this dry-stack retaining wall.**

The original wall **that ran along the driveway was unstable due to poor drainage and soil-retention problems.**

BEFORE YOU BEGIN

Replacing a large wall requires considerable staging. Before you begin, decide where you will put excavated soil, gravel, stone from the old wall, and stone for the new wall. Also, fragments from shaping stone can fly a long distance at a surprising speed. If you are working around vehicles or windows, be sure to cover them.

When a wall follows an asphalt driveway, the layout is a given. Sometimes trying to figure out how to lay out a wall isn't so simple; you may have to use stringlines (mason twine), batter boards, or garden hoses to lay out your project.

➔ **See "Layout Tools," p. 32; "From Plan to Ground," p. 78.**

When shopping at the stoneyard, choose stones that are uniform in size and close to rectangular. They are generally more expensive but require less chiseling. They also stack faster and make a stronger wall.

Calculating Materials **To determine how much stone you will need, divide the square-foot area of your wall by 25. For dry-stack applications, it generally takes about one ton of stone to make 25 sq. ft. of stonework. For example, if your wall is 50 ft. long and 3 ft. high, multiply 50 by 3 to get the area (150 sq. ft.). Then divide 150 by 25 to find how many tons of stone you'll need (6). It's smart to buy extra (10 to 15 percent, or 6½ to 7 tons in this example), in case there are some inferior stones you don't want to use.**

Even a small retaining wall is a project that requires a significant time commitment and a sizable staging area.

WHAT YOU'LL NEED

A considerable amount of material is required for a retaining wall, so set aside a large staging area.

- 1 ton fieldstone per 25 sq. ft. of wall
- 1 ton gravel per 30 sq. ft. of wall
- 1 roll (6 ft. x 50 ft.) filter fabric
- 4-in. perforated drainpipe for total length of wall
- Drainpipe connections
- Shovels
- Mattock
- Tamper
- Ax
- Wheelbarrow
- Levels
- Laser level
- Masonry tools
- Plywood and plastic for staging

WHAT CAN GO WRONG

Using too many shims can give a wall a loose and unorganized appearance. Instead, select and shape stones that fit well.

SETTING A DRY-STACK FOOTING

Set the footing for a retaining wall 8 in. below grade on compacted soil. Dig the footing trench 3 in. wider than the bottom of the wall ❶ and compact the soil with a hand tamper ❷. If you have to dig down deeper to reach undisturbed soil in any one area, compact the low spot, fill it with clean gravel, and then compact the gravel ❸.

→ **See "Retaining Walls," p. 9.**

Lay landscape fabric or filter fabric across the bottom of the trench and over the embankment ❹. Lay any extra filter fabric on top of the bank and secure it temporarily with stones ❺. Later, it will be folded under the capstones; wait until then to cut off the excess.

Lay rocks on the filter fabric to form a footing ❻. Bigger stones salvaged from the original wall can serve that purpose. If you are using new stones, choose large ones that would be difficult to shape. Spread gravel between the footing stones, and settle the stones by rapping them with a small sledgehammer or mallet ❼. Tamp the entire footing area to lock the stones and gravel in place ❽.

1 **Dig the footing trench deep enough to reach undisturbed soil; it should be 3 in. wider than the wall's base, in this case, 24 in.**

2 **Compact the trench bottom with a hand tamper. Pay special attention to the soil that will be under the wall's taller sections.**

3 Use gravel to fill areas that had to be dug deeper to reach undisturbed soil, and compact with a hand tamper.

4 Drape filter fabric across the trench bottom and up the bank. Wait until later to cut off the excess fabric.

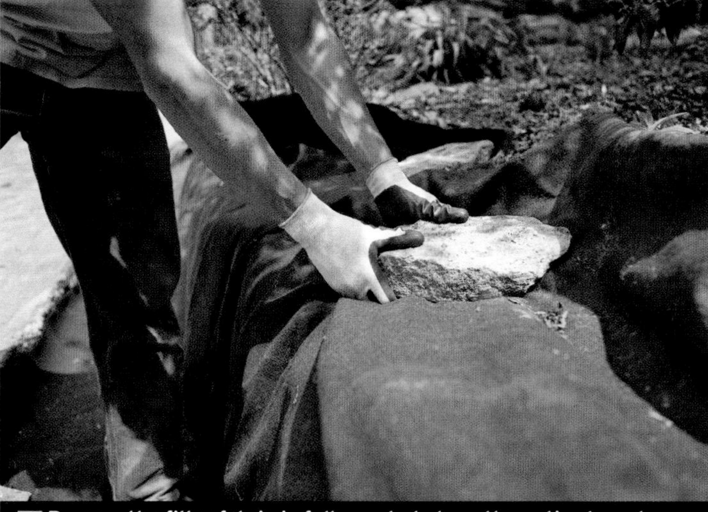
5 Be sure the filter fabric is fully seated along the entire trench width, and secure it with stones.

6 Lay footing stones on the trench bottom. Large stones that may be difficult to shape can be put to good use here.

7 Fill around the stones with gravel. Then strike them with a mallet to settle both the gravel and the stones.

8 Compact the entire footing with a hand tamper before laying the first course.

INSTALLING A DRAIN LINE

Walls built without mortar naturally let water that accumulates in the soil escape between the stones. However, water can also collect behind the footing. To ensure good drainage, you can install a perforated pipe behind the wall and cover it with filter fabric to prevent it (and the gravel beneath it) from clogging with silt.

Locate the drain as close as possible to the bottom of the footing and embed it in the gravel behind the wall stones ❶. Here, we are using a 4-in. single-wall perforated pipe. Surround the drain line with enough gravel to prevent the wall stones from crimping or crushing it as the wall settles ❷. Take care not to damage the pipe with a shovel as you spread gravel, and avoid setting stones directly over the pipe as you continue building the wall ❸. If the lowest point is at one end of the wall, direct the pipe to daylight where the wall ends, surround it with wall stones, and cut it flush to the wall ❹.

1 Lay the perforated drain line at the back of the footing and push gravel against it to hold it in place.

2 Surround the drain line with 2 in. to 3 in. of gravel after the first course is laid.

3 Continue building the wall around the drain line, but do not lay large stones on top of the perforated pipe.

4 Secure the pipe end at the grade's lowest point and cut it flush with the wall face.

ADDING A DRAIN-LINE T

If the lowest point is between the ends of the wall, add a drain-line T and run your drain line from it. First, use a transit or laser level ❶ to determine the lowest point along the footing. This is easy to do by sighting a common tape measure ❷.

➔ **See "Levels, lasers, and transits," p. 34.**

Lay drainpipe along the length of the wall, and mark the lowest spot ❸. You can use a small rock as a marker instead of scratching or painting the asphalt driveway; chalk is a good alternative. Use the T-fitting to determine where to cut the drain ❹. Remember to allow for overlap where the T-fitting and the drain line couple together. Cut the pipe and fasten it to the two sides of the drain line ❺.

Build the wall around the T-section. To make the drain outlet less obtrusive, recess the T-opening a few inches ❻. Position a flat rock under the outlet to direct the water away from the wall footing. Stack rocks on each side of the pipe, and bridge them with a horizontal stone to carry the weight of the wall ❼. Then continue building the wall over and around the pipe ❽. >> >> >>

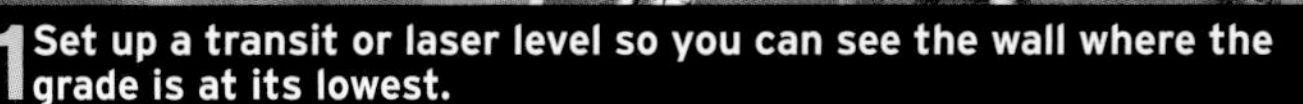

1 Set up a transit or laser level so you can see the wall where the grade is at its lowest.

2 Site a tape measure to determine the lowest point along the wall, which is the best place to put the drain.

3 Mark the lowest spot, and position the drain-line T so it's centered on the mark.

4 Use the T-fitting to mark where to cut the main line. A sharp utility knife is usually sufficient to make the cut.

ADDING A DRAIN-LINE T (CONTINUED)

5 Connect the T to the perforated drain, and check that it is still centered on the lowest point.

6 Extend the T with a short length of drain line. It should be slightly recessed from the wall face for the sake of appearance.

7 Use stones to create an opening with sides and a top to prevent the outlet pipe from being crushed by the weight of the stones.

8 Continue building the wall around the pipe. Span the drain opening for at least one more course to help carry the wall's weight.

SETTING WALL STONES

Before you start stacking stones, prepare the site by removing soil left from the excavation and set up a work area. Lay down plastic sheeting to protect driveways and patios ❶. Lay out three sheets of plywood as well; pile unshaped stones on one, pile gravel on the second, and reserve the third one for shaping stones. Keep the work zones close together to minimize unnecessary movement.

➔ **See "Establishing Staging Areas," p. 77.**

Use large stones, with the best edges facing out, for the bottom course; they will anchor the wall, both visually and structurally ❷. Vary the stone heights to break long-running horizontal joints. Also, cross (or break) all vertical joints when possible. This is important for both structural and aesthetic reasons. The more often you cross the joints, the better the stones lock together and the better it will look ❸.

Once you lay a row of 9 or 10 stones, backfill with clean gravel ❹. If any of the stones in the next course are wider than the previous course, the gravel will provide a level area large enough to accommodate them. Add enough gravel under stones that taper to ensure that the tops of the stones are level from back to front. This not only makes the wall stronger but also makes it easier to set the next row of stones. If a stone is not level from side to side, use the small shards left over from shaping stones to shim it ❺. Shims are also good for filling small voids that are inevitable regardless of how carefully you shape the stones. However, don't use shims as a substitute for shaping stones to fit. >> >> >>

1 Set up a zone where there is enough room to work, keep materials close at hand, and have unobstructed access to the wall area.

2 Lay the first course. Use larger stones to anchor the wall both visually and structurally.

3 Cross, or break, both horizontal and vertical joints. Breaking joints makes the wall stronger and more visually appealing.

SETTING WALL STONES (CONTINUED)

4 Fill behind set stones with gravel until it's flush with the top course; settle the gravel with a hammer handle's butt end.

5 Add shims to level stones and to minimize transitions of small gaps between stones.

6 Use a tape measure to help find stones that need as little work as possible.

7 Level the stone tops. Stones will sit more securely over time if they are laid on a level surface.

8 Establish a pitch, or batter, to the wall face. The slight backward tilt helps prevent the soil behind the wall from pushing it over.

9 Use levels and stringlines to create a uniform face. Sight along the wall often to check your work.

As you build the wall, keep a tape measure handy. Many novice wall builders try to judge size by eye alone. Using a tape measure also helps make it easier to see where the joints will cross ❻.

The other tool to keep on hand is a level. Periodically, check the coursework to keep the tops of the stones level ❼. Even on sloped ground, keep the coursework level. Another reason to use a level is to maintain a batter (a slight backward tilt to the face of the wall). A 1-in.-per-foot batter is good for typical walls ❽. Lastly, use the level to keep the face uniform ❾. Even for a curved wall, eyeing the wall with a level is a good way to identify stones that are set too far in or out.

For more, see "Make a Clean Break Using Chisel and Sledge," p. 24; "Shaping Stone," p. 196.

EFFICIENT STONEWORK

The key to shaping stones efficiently is to select the right ones. If you look at the photo at right, two stones pop out. The first is the rectangular stone, at lower left. This stone, with its even, parallel faces, is almost ready to place despite being a little narrow; it will not require much work. The big stone that I'm resting my right hand on is irregular in shape, and worse, has a big bulge on the bottom face. A bulge like this is very difficult to remove, so I would set this stone aside. The stone upon which I have my left hand is marginal; before I spend much time with it, I would try to remove the bulge across its bottom. The stone in the middle is probably the best of the lot, with its wide rectangular shape and enough depth to add stability to the wall. By sharpening your eye when selecting stones, you'll build more wall in less time.

SETTING CAPSTONES

The height of the retaining wall is largely determined by the height of the bank against which you're building. When the wall is a few inches below the bank, begin to anticipate how you will set the capstones.

Capstones will likely be of different thicknesses. Rather than trying to break them to the same thickness, adjust the top row of wall stones to accommodate the capstone variations. The goal, of course, is a flat, level row of capstones. You may have to test-fit stones several times before finding combinations that work ❶. The height of your wall should be 1 in. to 2 in. below the edge of the bank, and the capstones should pitch forward slightly to allow water to flow over the wall during watering or heavy rain.

When you're ready to set the capstones, fold the filter fabric to the wall, cut off any excess ❷, and cover the fabric with gravel ❸. Then set the capstones and backfill with soil ❹.

Capstones, in general, need only be deep enough to cover the top of the wall. However, periodically select a stone that spans from the front of the wall well into the bank ❺. This will help keep the wall stable over time. In addition, frequently sight along the wall or check it with a level to ensure that the top leading edges are true ❻.

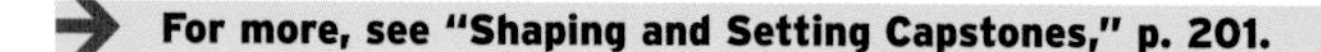
For more, see "Shaping and Setting Capstones," p. 201.

1 Test-fit capstones to see whether they are large enough and if the previous course is the correct height given the cap's thickness.

4 Capstones tend to be large; plan on having extra help on the day you plan to set them.

2 Fold filter fabric over the gravel; trim it where it meets the back of the wall.

3 Fill behind the wall stones with gravel until the gravel is flush with the top of the stones.

5 Fill behind the capstones with soil. Compact the soil to help it stay put until it is properly landscaped.

6 Align the top lead edge of the capstones. This helps give the wall an ordered and substantial appearance.

STONE PATIO

A STONE PATIO CAN BE LAID OVER an existing concrete patio, over a new concrete slab, or over open ground. Whatever the base, a stone patio can enhance your home and also provide a flat, functional living space in your yard. With the techniques you'll learn in this chapter, you can also build landings, door stoops, grilling platforms, and other areas that involve flat, mortared stonework.

For this project, we chose to lay Tennessee flagstone over an existing entrance patio. In this chapter, you'll learn how to apply an antifracture membrane, establish the proper slope with stringlines (mason twine), shape edge stones, lay and level stones in a mortar bed, and grout joints.

PRACTICAL CONSIDERATIONS

The best type of stone to use for a patio is flagstone. It's generally quarried in flat pieces that are easy to break with a hammer or cut with a saw. In many stoneyards, there are a variety of colors to choose from.

See "Make a Clean Break Using Chisel and Sledge," p. 24.

If the supplier is going to deliver the flagstone and unload it with a forklift, have the driver place the pallets as close to your work site as possible. If you are having the pallets dumped, you may not be able to get close to the work site, but do the best you can—and tell the driver to be careful when dumping the pallets to avoid breaking the larger stones. If your project is small and requires less than 1 ton of flagstone, unload the stone by hand to avoid breaking them or scratching the faces.

Unload sand and cement close to a water source for mixing cement and rinsing tools and equipment. If this is not possible, transport water in 5-gal. buckets to the work area. If you are laying flagstone on soil, make sure any fill is compacted and you use a good bedding material.

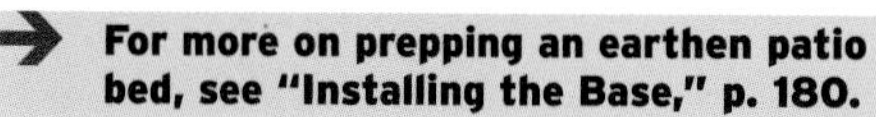
For more on prepping an earthen patio bed, see "Installing the Base," p. 180.

Whether you are installing flagstone on the ground, on a new slab, or on an existing slab, the first step is to make sure you have a solid surface on which to build. A mortared patio will most likely fail if installed on crumbly concrete or on a slab without proper pitch. In addition, applying an antifracture membrane on a concrete slab over the specified primer will go a long way toward preventing cracks that occur due to settling.

See "Barriers, Filters, and Membranes," p. 69.

Avoid installing flagstone (or any mortared masonry) if the temperature is expected to fall below freezing during the installation process. If the mortar doesn't create a strong bond before it freezes, it will be weak.

Flagstone is available in a variety of colors, thicknesses, and sizes. It's sold by the pallet, each of which has a tag indicating the tonnage.

If you are laying stone over an existing patio, it's a good idea to apply an antifracture membrane between the stone and patio to help prevent cracks from forming in the finished stonework.

BEFORE YOU BEGIN

Gather tools and materials and create a staging area as close to the work site as possible. Next, clean off the patio with a broom or flat hoe to remove any large pieces of debris. Use a gas- or an electric-powered blower to remove dust and dirt, followed by a wet mop or scrub brush. The cleaner you get the patio, the better the antifracture membrane will adhere.

It's always a good idea to protect finished surfaces when doing stonework. Spread plastic or canvas drop cloths over nearby lawn and garden beds. Tape edges of adjacent siding or other vertical surfaces, and cover windows and doors with plastic, plywood, or foam board. A little bit of protection will save a lot of headaches down the road.

See "Protecting the Surroundings," p. 75.

A well-organized staging area, close to the work site, **helps the project go smoothly. Here, plastic protects the driveway; plywood creates a space for mixing mortar; and bags of portland cement are elevated and protected from water.**

Clean the existing slab thoroughly. **A power washer is the ideal tool for the job and can be rented at a home improvement store. If a power blower isn't sufficient to remove dirt, it's worth the effort to clean it with water.**

WHAT YOU'LL NEED

- 1½ tons 2-in. flagstone (for a 120-sq.-ft. patio)
- 2 yd. masonry sand
- 12 bags portland cement
- Antifracture membrane with primer
- Paint roller with a long handle
- Wheelbarrow
- Shovel
- Tape measure
- Nylon mason twine
- Cowhide, cotton, or leather gloves
- Masonry tools
- Levels
- 5-gal. buckets
- Pencil

APPLYING ANTIFRACTURE MEMBRANE

Antifracture membranes vary by supplier but generally include a roll-on primer ❶ and self-adhesive membrane. It's always a good idea to read the installation instructions specific to your product.

Apply the membrane along the low end of the patio first ❷. If possible, have at least one helper. Overlap the membrane strips by 2 in. as you proceed up the patio ❸. The overlap allows water to flow over the seams rather than into them. Terminate the membrane at an adjacent wall by running it up the side by 2 in. ❹. Finally, cut away any excess membrane hanging over the patio edges.

Once the membrane has dried, it is safe to walk on. If you are going to store materials on the patio while you work, make sure you avoid puncturing the membrane. One good idea is to place sheets of plywood over the membrane to protect it while you are working.

There are a few things to remember at this stage. Use gloves when handling the primer and wear a respirator to avoid inhaling the fumes. The membrane is sensitive to extreme temperatures and will get gummy in warm weather. Use a utility knife rather than scissors to cut the membrane.

WHAT CAN GO WRONG

On warm days, membranes may be difficult to work with because they stick to themselves. They will also stick to rubber gloves, so wear cowhide, cotton, or leather gloves instead.

1 **After the slab is dry, apply primer to the concrete with a paint roller on a long handle.**

2 **Apply the first course of membrane at the low end of the patio and work your way to the high end.**

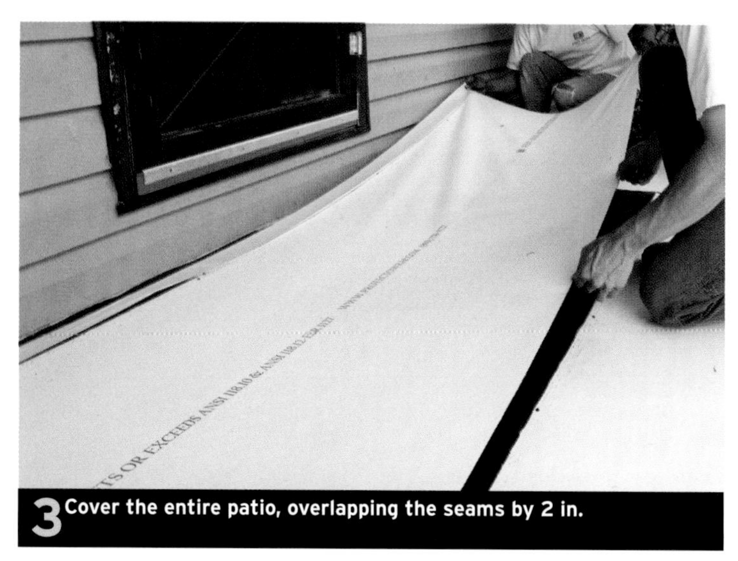

3 **Cover the entire patio, overlapping the seams by 2 in.**

4 **At the wall, fold the membrane up the side 2 in. and press it in place.**

LET TWINE BE YOUR GUIDE

A patio needs to slope to prevent pools of water from accumulating on the patio surface. Slope also directs water away from adjacent buildings. If you're covering an existing concrete patio, use a level to determine whether there is adequate slope, or pitch, and that it's in the right direction ❶.

A great way to make sure you create and maintain proper pitch of flagstone during installation is to establish stringlines with mason twine. The twine indicates the finished elevation before you begin. Not only do stringlines serve as a guide to keep the patio surface flat but they also help establish straight lines on the front and sides of the patio. Generally, nylon mason twine works best.

To set up a stringline, mark the highest point of the patio where it meets the wall of the house ❷. (In the case of this patio, it is a fixed point under the door threshold.) Drive two stakes, one on each side of the patio, and tie twine between them ❸. Use a line level to adjust the stringline so it intersects the patio's highest point and is level ❹. This line represents the high side of the finished patio.

The patio needs to slope away from the house 1⁄4 in. per foot, so calculate the necessary elevation change by multiplying the width of the patio in feet by 1⁄4 in. The patio shown here is 7 ft. wide, so it required a 1 3⁄4-in. drop from the high end to the low end. At the low side of the patio, mark the patio's finished top on a stake (or on adjacent structures) ❺. When measuring up from the existing patio, remember to account for the existing slab's pitch.

Tie a level stringline across the entire low side. In a property with two columns on the low side, you could tie twine to nails driven into the mortar ❻ (as here). For projects without columns, drive stakes adjacent to the patio ❼. Keep the stringline 1 in. away from the old patio's edge to allow for the flagstone's 1-in. overhang ❽.

Once you have the high and low sides defined by stringlines, begin setting the perimeter stones. You'll use these perimeter stones to set up stringlines on a grid, which is described in the next section. If you take the time to carefully establish stringlines, setting stones accurately becomes a lot easier. >> >> >>

ESTABLISH THE PITCH WITH STRINGLINES

A patio should slope away from the house. If your existing patio is level or has negative pitch, you'll have to pitch the flagstone to create the pitch change needed for drainage. The minimum drop for proper flow is 1⁄4 in. per ft. If you're topping an existing patio that's level or has a negative pitch, check to be sure you have enough room to achieve the required pitch. For example, a patio that extends 10 ft. from a house will need a total drop of 2 1⁄2 in.

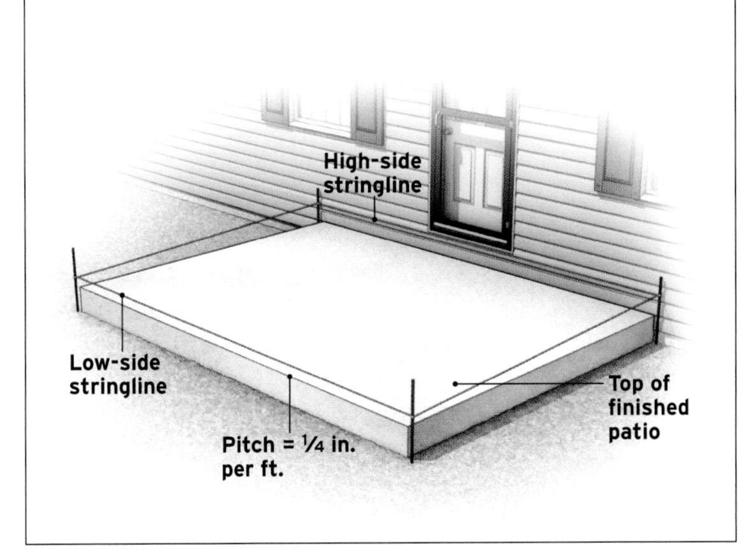

1 Determine the existing pitch by placing a mason level across the patio and raising one end until the bubble reads level.

2 Establish the patio's highest point at important transition points, such as adjacent steps, decks, and doorways.

3 Tie twine across the patio's high side. Use stakes to hold the twine if the siding prevents you from placing nails in the structure.

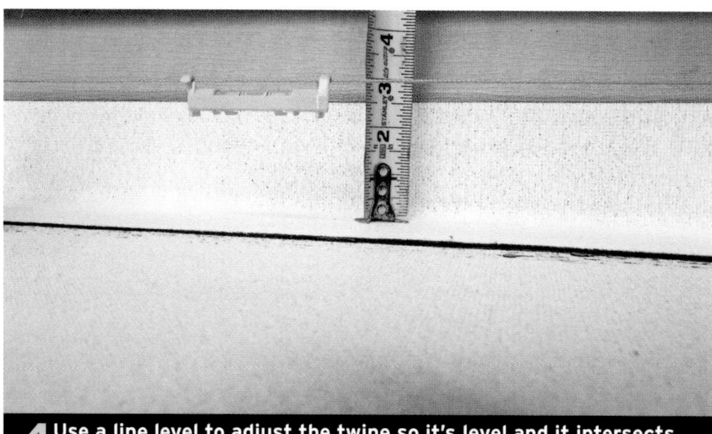

4 Use a line level to adjust the twine so it's level and it intersects the previously established high point.

5 Mark the elevation of the finished surface at the patio's low side after calculating the desired drop.

6 Drive nails or stakes at each side of the patio and tie the low-side twine to them.

7 Pull the twine tight and use a line level to adjust it. Reposition the stakes or nails if necessary.

8 Space the stringline far enough from the front edge to allow for a 1-in. overhang.

TRADE SECRET

A quick way to determine a patio's pitch is to dump a bucket of water on the surface and see which way and how fast it flows. If the water races away from the house, you may be good to go.

SETTING PERIMETER STONES

Once you've established the patio's pitch with the high and low stringlines, you're ready to mix a batch of dry mortar and start setting perimeter stones. Space stones 4 ft. apart in a grid pattern. Later you'll run stringlines over these stones to help keep the patio on the same pitch plane.

➔ **See "Mortar and Concrete," p. 61.**

Begin with a stone under the high-side stringline. Spread enough mortar so the stone sits 1/2 in. above the twine ❶. Pack the mortar around the stone base with a trowel to ensure that it fully supports the stone. Do not leave any gaps ❷. Strike the stone with a rubber mallet to set it level to the stringline ❸. When a stone is set correctly, you can see only a sliver of light between it and the twine ❹. Avoid striking the stones too hard; they break easily.

Set the next stone under the low-side stringline. Then run a twine over the high- and low-side stones ❺. There are a few different ways to secure the twine. If the stone is next to a wall, embed the end of the twine in the mortar under the stone ❻. Securing the twine this way anchors it enough for tightening but allows easy release. To secure twine over an edge stone, wrap it around a rock, and pull the twine tight ❼. >> >> >>

WHAT CAN GO WRONG

Working with mortar that is too wet or too dry will cause a variety of problems. If the mix is too wet, the stones will float, or move, after you set them. You'll find it difficult to maintain joint spacing and keep the stones level. If the mix is too dry, the stone won't set or sink into the mortar. You'll have a hard time adjusting the stone because it simply won't move very much with each hammer blow.

1 Spread mortar and place the first stone. The stone should push the twine up 1/2 in. before it is set.

2 Press mortar around the base of the stone with a trowel so the stone is fully supported.

3 Set the stone with a rubber mallet. With a few blows of the mallet, the stone should set into the mortar 1/2 in.

SETTING PERIMETER STONES (CONTINUED)

Twine strung from the high side to the low side of the patio begins to form a grid you can use to set the rest of the stones. By setting stones to this line, you'll maintain the overall pitch of the patio. As you progress, use a combination of the stringlines and levels to ensure that all the stones are set to the same plane ❽. Take the time to set all the stones to the right elevation and pitch, even if it takes several tries.

4 The gap between the twine and the stone should be as small as possible; they should not touch.

5 Set a stone aligned with the low-side stringline in the same manner and run twine over the tops of the low- and high-side stone.

6 Secure the twine at one end by passing it under the stone so it is embedded in the mortar. Do not tie the twine to the stone.

7 Tie a rock to the twine's other end, and pull it tight over the stones to establish a guideline that shows the patio's pitch.

8 Continue to set stones using the twine, levels, and straightedges so they align with the pitch.

SETTING EDGE STONES

A straight border, or edge, defines the patio and makes it look neat. Always set edge stones to a stringline. In many cases, the low-side stringline can also serve as your edge line. If not, run a second line to which you can set the stone's top leading edge. Align the stringline with the overall pitch plane of the patio, and allow for a 1-in. overhang for the edge stones.

Select a stone that already has one fairly straight side ❶. Draw a cut line using a straightedge. Then trim the stone as near to the line as possible ❷. Be patient: Removing small chips with each hammer strike reduces the risk of the stone breaking in the wrong spot. When one side is finished, mark and trim the same edge from the other side. When you like how the leading edge looks, dry-fit the stone and mark a trim line on the adjacent edge stone. Aim for an even 1/2-in. to 1-in. gap between the stones ❸. Also, it's a good idea to shape the back edge so subsequent stones are easy to fit ❹.

See "Fitting Stones," p. 110.

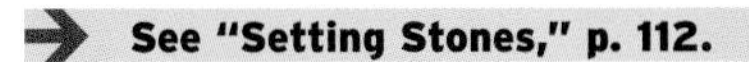
See "Setting Stones," p. 112.

1 Select a stone with an edge that is already close to straight. Then use a square to mark a line as close to the edge as possible.

2 Working carefully, use a brick hammer to trim the edge of the stone to the line.

3 Dry-fit the stone and mark the trim line to the adjacent stone.

SETTING EDGE STONES (CONTINUED)

Lay an even bed of mortar under the entire stone ❺. Use the back of your hand to lightly compress mortar at the edge and help it stay put ❻. As with the other stones, when you first place the stone on the mortar bed, it should be about ½ in. above its final height. Align the stone to the stringline and set it with a rubber mallet ❼. With a pointing trowel, pack mortar under the stone at the patio's edge ❽.

It's difficult to finish even a small patio in a single day, so before you leave a job that's on pause, make sure you rake all the mortar out of the joints and cut away mortar from the side of the stones. By doing so, you'll have a clean area in which to work the next day and won't have to contend with mortar that has set in the wrong place ❾.

➔ See "Setting Tools," p. 26.

WHAT CAN GO WRONG

Setting a stone under a stringline that's even a little bit too high will have a ripple effect. If the twine is pushed up in one spot by a high stone, you'll end up setting the adjacent stones too high as well. The end result will be a hump in the patio.

4 Mark and trim irregularities at the back of the stone so the stones you set behind it are easier to fit.

7 Align the stone with the stringline, leaving the desired gap between adjacent stones. Then set the stone.

5 Spread an even bed of mortar, covering the entire area under the stone.

6 Use the back of your hand to lightly compact the mortar near the edge so that it will not fall away as you set the edge stone.

8 Pack mortar under the edge with a pointing trowel. If the mortar won't pack well, check its moisture content.

9 Remove excess mortar at the end of the day. Any hardened mortar will have to be chipped out before you can continue.

FITTING STONES

Setting the stones properly in mortar will ensure that your flagstone patio lasts a long time, but the way you fit the stones together will determine whether the patio looks good. A useful rule is to leave a ½-in. to 1-in. gap between the stones. You can go narrower, but keep in mind that anything less than ¼ in. will make it hard to press grout into the joint. Being consistent with the joint width will give you a cleaner, more professional-looking pattern. With large stones, a 1-in. joint may look good, but I would stay away from joints wider than 1 in. unless you plan to plant grass or a groundcover between the stones.

You can shape and set each stone as you lay them. However, if this is your first patio, it's better to shape several stones at once and dry-fit them on the ground before mixing a batch of mortar. This will give you more time to make the necessary marks and cuts. Clear out a 3-ft. by 3-ft. section beside the patio and practice. If you spread mortar and then try to cut the rock, you're less likely to do a good job simply because you are hurrying to set stones before the mortar dries.

Shaping a stone

When shaping a stone, complete the fit of one edge before marking and shaping a second edge. To do this, lay the stone over the spot where you want it to fit and mark an approximate cut line with a pencil ❶. Shape the stone to the line, checking its fit periodically ❷. Once the first edge fits, move on to the next edge, marking and shaping it in the same manner ❸. Finally, consider how your next stone will fit up against it ❹. Try to create an edge that will make it easy to fit the next stone by straightening edges and removing irregularities.

Shaping stones in this way is like putting together a jigsaw puzzle, except that you have to cut the pieces as you need them. If you develop a good fitting technique and use a little patience, you will soon become faster and more efficient ❺.

For more, see "Dress a Corner Stone with Hammer and Chisel," p. 25.

TRADE SECRET

Use tile spacers to keep the joint widths consistent. Spacers keep you from having to measure the joints repeatedly. For best results, use ½-in.- to 1-in.-wide spacers.

1 Lay the stone over the space and mark the first cut line using the adjacent stone as a guide.

4 Prepare an edge for the next stone by removing irregularities.

2 Trim the stone to the line. For a more controlled process, undercut the edge, then slowly chip the top surface back to the line.

3 Repeat the process of marking and shaping each edge adjacent to neighboring stones.

5 Plan ahead so that the last stone you place is a comparatively smaller stone in the middle.

WHAT TO AVOID

If you want your patio to look professionally laid, avoid these three common mistakes: sloppy joint work, such as joint widths that are too wide or inconsistent; misaligned stone tops; and mortar stains on the stone faces.

SETTING STONES

Setting stones in mortar is the same whether the stone is in the middle of the patio or at the edge. First, create an even mortar bed that supports the entire stone ❶. Between stones, create a trough in the mortar, which gives excess mortar a place to go as the stone is set ❷. If a stone doesn't set right (too high, too low, or otherwise off-kilter), remove the stone, fix the mortar bed, and reset the stone ❸. After setting each stone, use a pointing trowel to remove excess mortar from between the stones ❹.

➔ **See "Setting Tools," p. 26.**

Once you finish setting all the stones, the mortar will need at least one day to cure. As you build out the flagstones, remember not to strand tools where you cannot reach them. Similarly, remove mortar from the joints so they will be ready for grout when the time comes. Another handy tip: If rain is in the forecast, cover the patio with plastic sheeting.

1 Spread an even bed of mortar under the entire stone.

TRADE SECRET

If it's a warm day, wet the back of the stone before setting it in mortar. This helps create a stronger bond and prevents the stone from wicking moisture from the mortar.

2 Channel the mortar at adjacent stones to provide a space for the mortar to move to when the stone is set.

3 If the stone is setting too high or too low, remove the stone and adjust the mortar bed.

4 Remove excess mortar from between the stones before the mortar sets to allow space for grout.

FLOATING STONES

"Floating" is a term used when the mortar bed is too wet and flagstones that are already set move as you set adjacent stones. For example, the pressure of setting one stone can cause adjacent stones to rise. When the mortar dries, none of the stones sits level. You should be able to tell if the mortar is too wet right away. If it takes a few swings of the mallet to set the first stone, it is probably right. If you can push the stone down with your hand, the mortar is too wet. To stiffen wet mortar, add equal parts sand and portland cement.

DRY-MIX GROUTING

Before you begin grouting, if you haven't already done so, mask any surfaces adjacent to the patio that need protection, including siding, woodwork, or other stonework ❶. The bottom edge of the masking tape should be level with the top of your grout line. Also double-check the space between the stones and remove unwanted mortar you may have overlooked. If there is mortar filling the joint and it is difficult to remove, loosen it with a sharp brick hammer or cold chisel. If this doesn't work, use a small grinder or saw fitted with a masonry-cutting blade. Finally, remove all debris, including dirt, dust, and loose cement pieces, from the joint with a broom or blower ❷.

Mix a batch of grout and transfer it to a 5-gal. bucket. It is much easier to grout on your knees, so you'll want a pair of kneepads. Hold a handful of grout next to the joint and use a pointing trowel to pack the grout tightly into the joint ❸. Leave the grout raised slightly above the stone; later you will scrape it flush. If one stone is slightly above an adjacent stone, make sure the grout comes all the way up to the top edge of the stone so there will be nowhere for someone to catch a toe and trip. Use your thumb to press firmly on the joint every now and then to make sure you are packing the grout tightly enough. At open edges, such as the outer end of a patio, hold your hand under the joint and pack it with grout using the pointing trowel ❹.

See also "Dry-Grouting Flagstones and Capstones," pp. 30–31.

Let the grout dry for an hour or two. Then use a pointing trowel to scrape it flush with the top of your stones ❺. If the grout seems wet and soft, let it dry a bit longer. If you are grouting during hot weather, consider scraping off the excess grout sooner.

Last, scrape the joints along the edge ❻. Sweep up or blow off the excess grout fragments when you are done. Thoroughly clean the finished surface using a masonry detergent to remove tough stains. After the patio is clean and dry, apply a sealer ❼.

1 **Mask adjacent siding, decking, or other stonework before applying grout to prevent cement stains.**

4 **Hold your hand under joints at the edge to support the grout while you pack the joint.**

7 **After the grout has dried, thoroughly clean the patio and then apply a sealer.**

2 Use a power blower to remove all dirt and dust from between the stones before grouting.

3 Pack the joints full of grout. Use a pointing trowel to move grout from your hand into the joint.

5 Scrape the grout flush to the flagstone surface. Sweep up the excess grout as you work.

6 Flush up the joints at the edge and below the overhang.

USING A GROUT BAG

A lot of masons prefer to use grout bags rather than dry-mix grout. A grout bag has a narrow tip that allows grout to be precisely laid in the joint; it's a bit like decorating a cake with a pastry bag. Wet grout also provides a better bond between the grout and stone. However, you have to be very careful with wet grout. For it to come out of the bag properly, it has to be very wet, which can make a mess on the stones if you are not careful. The other disadvantage is that it takes a lot longer for wet grout to set. If temperatures are cool, it might take half a day, so plan accordingly.

To begin grouting with wet grout, prepare a grout bag. First fill the bag half-full ❶ with wet-mix grout. Then twist the bag's open end to close it off ❷ and fold it so that when the bag is squeezed the grout flows out of the nozzle ❸.

As with dry grout, make sure the wet grout is applied so it's slightly above the surface of the stones ❹. When the grout begins to set, scrape it level with the stones. Allow the cement that smears on the stones to dry. Then use a metal brush to scrub it off. After you have finished, use a sponge to wipe the top of the joint. Rinse and use the sponge to wipe the remaining cement off the stones. You may have to do this several times to get it all off.

WHAT TO AVOID

Messy or inconsistent grouting will reduce the appeal of your stonework no matter how hard you work at shaping and fitting stones. Using different amounts of cement and sand from one batch to another will create slight mortar color variations and make the stonework look patchy. When working with porous stones, consider using dry grout; it is much easier to clean off stones if they should become stained.

A MORTAR FOOTING

The process for building a flagstone patio on the ground is similar to setting one on a slab, except that it requires a thicker, 4-in. mortar bed. There is no guarantee that this type of installation will prevent the patio from cracking, but there are precautions you can take. First, excavate 8 in. to 10 in. down with a mattock and shovel, removing any roots or debris you find beneath the soil. This will allow enough depth for clean gravel, a mortar bed, and flagstone. Use a tamper or a plate compactor to pack the soil. Spread 3 in. to 4 in. of clean gravel over the area and tamp again.

Make the setting mortar the same way you made the mortar for the patio on a slab, but just make more of it. Set wire mesh in the middle of the mortar bed.

➔ **See "Wire Mesh and Lath," p. 66.**

Use this approach only if you are building a patio of 150 sq. ft. or less. For larger patios, install a concrete slab for support.

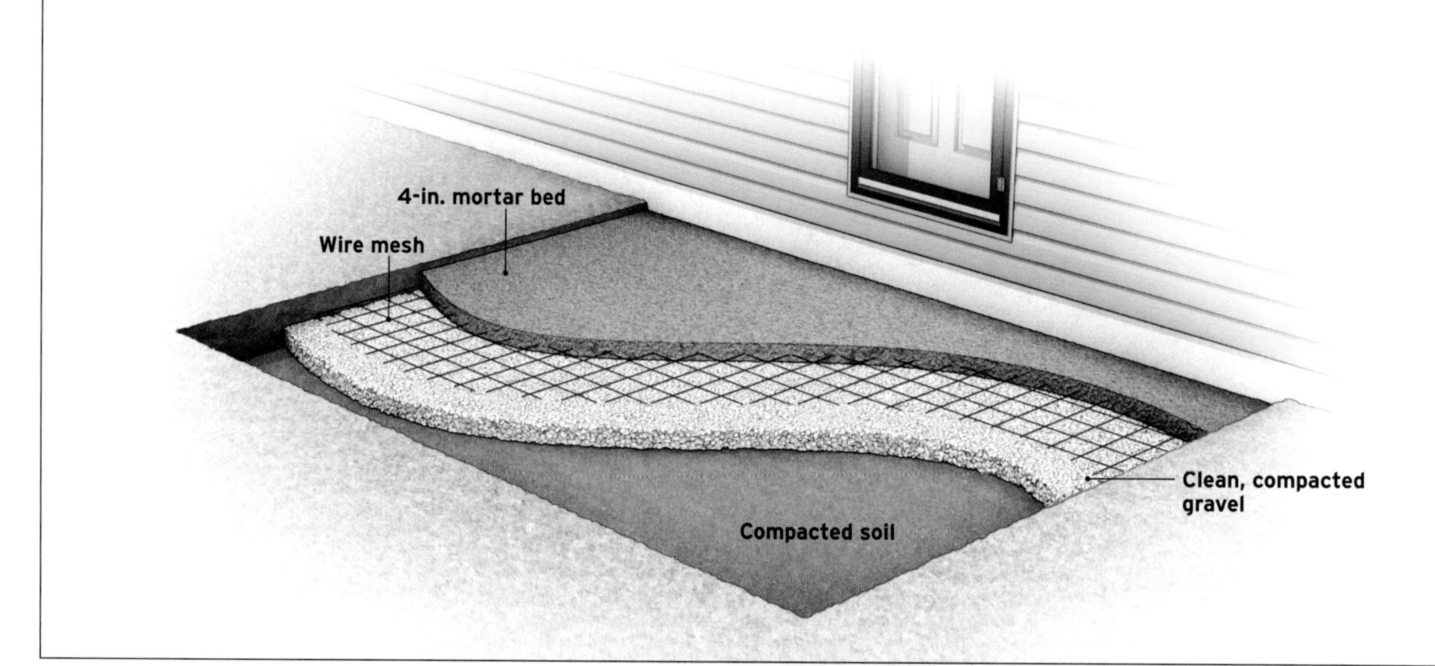

1 Load the grout bag half-full of a wet-mix grout.

2 Fold and twist the top to close it off.

3 Fold the top down to direct the flow out of the nozzle.

4 Squeeze the bag to grout the joints.

BRICK VENEER

BRICK VENEER IS A POPULAR siding choice, and the reasons are clear. Its look is timeless, it is durable, and it can significantly increase the curb appeal of your house. The whole topic of how to apply brick veneer could easily fill a book. However, many smaller brick projects can easily be tackled by learning the fundamentals of applying brick veneer.

For the siding project shown here, we installed a brick-veneer patch beneath a new window. In this chapter, we'll cover wall preparation, installing flashing, mixing brick mortar, weaving new brick with old brick, cutting brick, and creating a window sill (also called a brick rowlock). The method of laying brick veneer shown in this chapter can also be applied to veneering columns, fireplaces, and retaining walls. We conclude the chapter with a brick-veneered step project.

GENERAL PREPARATION

PREPARING THE WALL

LAYING BRICK

LAYING A WINDOW SILL

BRICK-VENEERED STEPS

PRACTICAL CONSIDERATIONS

Before you can begin a patching project, such as this one, you'll need to do some legwork to match the existing brick and mortar. Otherwise, the finished look may not be what you envision. In addition, consider the scope of the project and decide whether you need more help, equipment, or tools to complete the job in the time you have budgeted.

Matching bricks and mortar

A major question when taking on a brick-patching project is whether the brick can be matched. If the house is relatively new, the original builder may be able to help identify the brick manufacturer or supplier. If information is not available, take samples to local masonry suppliers and see if they can help you find a match. Be aware that matching brick is not an exact science, and you may never find a perfect match.

Matching mortar is less about chance than it is about trial and error. The color of sand, ratios of ingredients, and color additives all affect appearance. The best way to match mortar is to make up several sample mortared joints, keeping track of the recipes used. After the samples are fully dry, compare them to the existing mortar. This process may take some time, so try to settle on the right recipe well in advance of the project start date.

➔ **For more on brick, see p. 57.**

➔ **For a mortar recipe for brick, see p. 61.**

Matching mortar can be tricky with brick. The red mortar shown here would require a substantial amount of dye to achieve a proper match. Mismatched mortar would stand out and be unsightly.

Project size, scaffolding, and tools

If your brick project is bigger than the siding patch shown in this chapter, consider hiring an experienced bricklayer to help out. Even if you want to do some of the work yourself, working beside a veteran can be more informative than reading an entire book on the subject. You'll need to find a bricklayer who is open to teaching you; if you can, this is the fastest way to learn and could save you significant time and money in the long run.

Masonry work cannot be done from ladders, so any job that extends more than 4 ft. off the ground needs to be completed with scaffolding.

Typically brickwork cannot be done from ladders, so if your project is more than chest height off the ground, scaffolding is required. Rent scaffolding from your local equipment-rental supplier; you'll need it for the duration of the job.

 See "Scaffolding," p. 46.

If the project requires precisely cut brick, you'll also have to rent or buy a chopsaw and fit it with a diamond masonry blade. For small projects, you can get away with a circular saw fitted with a masonry blade. Aside from a chopsaw, a good set of setting and pointing tools will get you through most brick projects.

See "Cutting Tools," p. 38.

A small wet saw with a sliding table makes mitered cuts easy and relatively dust-free.

BEFORE YOU BEGIN

As always, establish a staging area close to the work zone, and gather all the tools and materials ahead of time. Staging for brick projects requires a much smaller area than for stonework because bricks come neatly stacked on a pallet. In addition, you don't need an area to shape them as you do for stone. You will need an area to mix mortar, preferably on a piece of plywood. You'll also need to establish a place to cut the brick that is away from open windows and intakes for air conditioners or air handlers. Dust clouds generated by cutting bricks with a dry blade can be voluminous.

See "Protecting the Surroundings," p. 75.

Because brick cladding is a defense against rain, have a water-management plan in place before beginning demolition. This plan should include a contingency for rain and other inclement weather during the project. Also, you'll need to have a plan should you discover rot or mold in the wood structure. Lastly, if the original brick is structural, consult an architect or engineer before beginning the project.

WHAT YOU'LL NEED

- Bricks for 25 sq. ft.
- 1/8 yd. masonry sand
- 5 bags Type S masonry cement
- Wall ties
- Weep inserts
- Roll asphalt flashing
- Mortar net
- Moisture barrier
- Masonry tools
- Saw
- Grapevine jointer
- 5-gal. buckets

FLASHING, FASTENERS, AND WALL PREP

Before you get to actually laying the new brick, any old brick (if there is any) needs to be removed, and the wall needs to be protected against moisture infiltration. The easiest way to remove bricks is to cut them along a joint with a masonry blade. With vertical cuts, this leaves a half-brick at every other row. To remove the half-bricks, cut the horizontal mortar joints between each brick ❶. Then strike the vertical mortar joint sharply with a brick hammer's claw ❷. It may take a couple of strikes to crack the mortar. Working in the vertical joint only, pry out the half-brick; then chip away old mortar. Once the old bricks and mortar are removed, you're ready to weave in new bricks with the old.

First, clean up debris and dust, and then flash the bottom of the opening with self-adhesive flashing membrane ❸. Tuck the flashing under the housewrap and let it lap over the footing to within 1 in. of the wall. This will allow water that gets behind the brick (and there will be some) to exit at the weep holes in the bottom row of bricks.

Brick veneers are required to have an air space between the back of the brick and the wall to allow water to escape. To keep the mortar from falling behind the veneer and clogging the weep and drainage system, place a drainage net behind the bricks, at the base of the wall ❹.

➔ **See "Barriers, Filters, and Membranes," p. 69.**

The final bit of wall prep, attaching wall ties, can be done before setting brick or as you progress. Wall ties are strips of corrugated metal attached to the wall at least every 16 in. and embedded in the brick mortar ❺. They prevent the veneer from sagging toward or away from the structural wall. Wall ties are used for both brick and stone veneers.

➔ **See "Wall Ties and Anchors," p. 67.**

1 Cut the horizontal joints slightly past the nearest vertical joint of the half-brick you intend to remove.

4 Install a drainage net at the bottom of the wall behind the bricks to maintain a free flow of water and air.

2 Break the vertical joint with a brick hammer's sharp end and pry the waste brick out. Then chip away the old mortar.

3 Flash over the sill to direct the water out of the weep holes and away from the interior walls.

5 Fasten wall ties to the framing at least every 16 in. to secure the veneer and to maintain the proper ventilating space.

LAYING THE FIRST COURSE

When patching to existing brickwork, the first course of brick often needs to be cut so it will align with the existing horizontal mortar joints. Measure from the footing to the top of the first course of existing brick. Allow for 1/2 in. of mortar under the brick, and make your cuts ❶. This cut is forgiving, plus or minus 1/8 in., so you can use a gas-powered circular saw fitted with a masonry blade. You could also use a standard circular saw or chopsaw fitted with masonry blades ❷. For less dust and more controlled cuts, use a wet saw with a sliding table.

See "Cutting Tools," p. 38.

To align the brick tops with the existing horizontal joint, drive a concrete nail into the joint on each side and pull a stringline (mason twine) between the nails to serve as a guide ❸. Be sure the string aligns with the top of the brick, not the middle of the joint.

Spread the mortar along the footing ledge and set the first few bricks, aligning them with (but not touching) the stringline ❹. Try not to spread too much mortar until you get the hang of it. For the first row of bricks on the footing, scrape the mortar joint flush, discarding mortar with a downward stroke ❺. Doing so wastes a little mortar but prevents dirt and loose gravel from contaminating the rest of the mortar you've mixed. Every third or fourth brick, insert a weep insert between the bricks ❻. Weep inserts are honeycomb plastic vents that replace a joint and allow moisture to escape from behind the bricks.

See "Barriers, Filters, and Membranes," p. 69.

When you're ready to begin the next row, move the stringline up one course and repeat the process.

TRADE SECRET

To make quick work of marking the cut line on bricks, hold the tape at the measurement between your thumb and forefinger. Hold a pencil against the tape's hooked end. Lightly press your forefinger against the brick's face and slide it along the bricks, marking as you go.

GETTING THE RIGHT COLOR MORTAR

Color-matching mortar can be difficult because nowhere during the process of adding dye and mixing mortar does the color look anything like the finished product. Testing various mortar and dye combinations is the only way to reliably achieve a match. If your patch is in a highly visible area, take the time to make several mortar samples and let them cure 24 hours; then compare the color to the existing mortar joints.

1 Measure from the top of the first row of bricks (indicated by the stringline) to determine the cut-line height.

2 Cut the bricks using a circular saw fitted with a masonry blade.

3 Drive masonry nails into the mortar joint above the first course, at both ends. Tie twine to the nails to serve as a reference.

4 Spread mortar along the sill and set the first row of bricks. Align the bricks' top edge with the stringline.

5 Scrape excess mortar using a downward stroke after setting the first row.

6 Add weep inserts between bricks to allow water to drain out from behind the wall.

LAYING AND WEAVING BRICKS

Good bricklayers are efficient with mortar and make throwing it look easy. Inexperienced bricklayers typically make a mess. However, there are some easy-to-master techniques that can make most novices look more like experienced pros than beginners. After you spread a layer of mortar, use the point of your trowel to rake the center of the mortar bed and form a V-shaped trough ❶. This creates a seat in which to set the brick. If you're using hollow brick, it pushes mortar down into the cells of the bricks to create a stronger joint.

After setting each brick into the mortar bed, hold the trowel at an angle, and scrape away the excess mortar ❷. Instead of throwing that mortar back into the bucket, use it to butter the end of the brick ❸ you just placed in the mortar bed. This is faster than buttering the end of each brick before setting it.

Once you get skilled at this process, you will develop a nice rhythm of setting a brick and scraping the mortar as if it were all happening in one motion. One hand is placing a brick while the other is swiping and throwing mortar. You'll find that there is a direct relationship between flipping your wrist and controlling where you throw the mortar. As you encounter wall ties, embed them in the mortar joint; apply mortar below and above the tie ❹.

When weaving bricks, place mortar under the brick and butter the end—but don't place mortar on the top ❺. Slide the brick into place end first (to minimize mortar mess on the brick face), gently prying against the existing brick with the trowel end ❻. Let the mortar set up for about 30 minutes. Then, using a trowel as a spatula, press mortar into the joint with a pointing trowel ❼. If the spacing doesn't work out perfectly, cut the bricks to fit. Sometimes, to avoid crossing joints, you'll have to cut two adjacent bricks by one quarter instead of cutting one brick in half ❽.

It is important to shape the joints after the mortar has stiffened so it won't smear, but don't wait until it is too hard to give it a profile. There is a fine line between when the jointer rakes the joints easily and smoothly and when it doesn't. One useful tool to help with this task is a grapevine pointer, which can sometimes match existing joints ❾. Experiment on the first course to find the right amount of time to wait. If you find that some of the joints don't have enough mortar, put a little on the edge of your trowel and pack it into the joint ❿.

1 Carve a V-trough after you spread mortar to help control the mortar flow and minimize brick movement when setting bricks.

TRADE SECRET

Often a nail won't set in the mortar joint exactly where you want it. To adjust the stringline to the top edge of the brick, place a chunk of brick, flat side down, on top of the string.

WHAT CAN GO WRONG

If it's a hot day, the mortar will begin to dry out quickly. Dry mortar is harder to throw, sets more quickly, and may not form as strong a bond to the bricks. Periodically check the mortar consistency and mix in a little water as necessary.

2 Scrape excess mortar frequently to help prevent staining the brick and for a neater job in general.

3 Use excess mortar to butter brick ends. This saves time and extra motion.

4 Spread mortar both above and below wall ties to fully embed them in the joint.

5 Butter only the bottom and end when weaving bricks. Slide the brick into place end first.

6 Pry the brick into position by gently levering the trowel's tip against bricks that are already set.

7 Use a pointing trowel and flat trowel together to press mortar into the joint above the brick.

8 Cut bricks to lengths that avoid alignment (or "railroading") of vertical joints when weaving into an existing wall.

9 Use a shaped pointing tool when the mortar has set slightly.

10 If there is not enough mortar between joints, add a little with a pointing trowel.

CREATING ROWLOCK CAP

A rowlock course is one where the bricks are laid perpendicular to the wall and overhang the veneer by 1 in. A rowlock course adds a nice cap detail to the veneer, and it accentuates the bottom of the window. It also provides a sill that sheds water. To determine at what point below the window to stop the last course of brick, hold a brick in position under the window frame ❶. Use a level to eyeball a minimum ½-in. pitch and align the underside of the brick with the nearest horizontal brick joint ❷.

Stop the typical coursework at this joint and clean off the top of the last course of bricks. Install flashing under the rowlock course in the same manner as over the footing. Adhere it across the top brick course starting 1 in. back from the edge. Then adhere it to the wall under the window frame ❸. If possible, try to locate the top edge behind or under existing window flashing.

To fit the brick and achieve the proper overhang, it's often necessary to trim the brick's back corner. With a little practice you can trim brick with a brick hammer rather than a saw. To do this, hit the brick where you want it to break. Then hit the brick on the opposite side, in the same place. Repeat this a few times, and the brick should break at that spot ❹.

Lay the outside bricks first ❺, placing a weep insert next to the brick flush with the veneer ❻. To make a straight rowlock, pull a stringline between the two outside bricks and align it with the bricks' top edge ❼. Spread a layer of mortar along the ledge and on top of the flashing, just as you did on the footing ❽. Incorporate the honeycomb weeps between the bricks every four or five bricks. As you lay each brick, butter the back and the sides ❾, which will ensure that mortar fills the joints when you push the bricks together and tap them into place ❿.

1 Position a brick under the window with at least a ½-in. drop from front to back.

To ensure that the brick spacing will have even mortar joints, as the last bricks are set, measure four or five bricks along the cap ⓫. Compare that measurement with the space left to fill ⓬ and adjust the joint thickness if necessary.

For the last brick, butter three sides of the brick before you wedge it in. If you can't get all the mortar in the joints, add mortar with a pointing trowel after placing the brick. Finish the joints with the pointing tool and remove excess mortar ⓭.

➔ **See "Pointers and Jointers," p. 28.**

>> >> >>

TRADE SECRET

To help prevent water infiltration, smooth the mortar joint where the rowlock cap meets the wall. Smoothing the joint makes it less porous.

WHAT CAN GO WRONG

Don't be left with space for only half a brick. Add the measurement of a brick to the mortar joint width, and then divide the overall gap width by that amount. Adjust the mortar-joint width until the number divides evenly (with no remainder).

2 Mark the underside of the brick and locate the joint nearest to where regular coursework will stop.

3 Install self-adhesive flashing to the bricks under the window when you have finished up to the rowlock.

4 Break or cut the brick's back corner to allow it to set under the window with the proper overhang.

5 Spread mortar and place the first bricks at each end. Press the bricks up under the window, but leave a 1/2-in. mortar joint below.

6 Place a weep insert against the bricks you've just set, flush with the veneer below.

7 Pull a string between the two bricks. Here, we used a pointing trowel to hold the twine; you can also use nails.

CREATING ROWLOCK CAP (CONTINUED)

8 To lay the rowlock cap bricks, spread mortar, slip the bricks under the window, and then press them down into the mortar.

9 Butter rowlock bricks along three edges, as shown (the unbuttered edge faces the sill). Avoid over-buttering, to prevent squeeze-out.

10 Set the bricks by tapping on the lower front face and lifting the back of the brick at the same time.

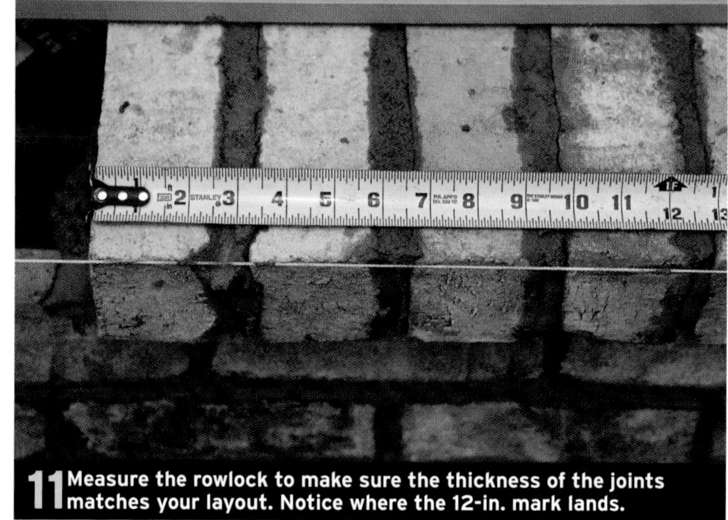

11 Measure the rowlock to make sure the thickness of the joints matches your layout. Notice where the 12-in. mark lands.

12 Measure the space for the remaining bricks. Even accounting for the joint space, the layout is off and will need to be adjusted.

13 Use a pointing tool to shape the mortar joints.

LAYING BRICK VENEER ON STEPS

In addition to siding, brick veneer is also an excellent choice for a host of other applications. A common one is to use brick veneer to cover the concrete block risers of outdoor steps. They offer a nice contrast to stone treads.

➔ **For more on installing concrete block for stone or brick-veneered steps, see "Making a Mock-Up," p. 142, and "Laying Block," pp. 143–144.**

A typical brick-veneer face is installed 5 in. from the block face, but for our steps the layout allowed only 4 in. To establish the riser location, measure the distance from the block face at both sides of the step ❶. Connect these marks with a chalkline. Repeat the process for each step and for the side of the block structure if you are going to veneer that as well ❷.

See "Hand-Mixing Mortar," p. 62–65.

Mix brick mortar on a piece of plywood or in a mortar mixer. You will need 18 shovels of sand and one bag of Type S masonry cement. It's a good idea to start installing risers at the top of the stairs, so you don't risk stepping on your finished work before it sets up. If you do start at the top, risers will extend behind the tread, not rest on top of it. If you prefer to start installing risers at the bottom, see "Setting Risers," p. 147-148.

Setting the bricks

To locate the first brick, mark the centerline of the stairs ❸. Then spread mortar inside the line on the top step, and set the first brick to the centerline ❹. As you continue setting the first row, spread a 3⁄4-in. to 1-in. mortar bed. This will leave a 3⁄8-in. to 1⁄2-in. joint after you set the brick.

Set the second row in the same manner as the first, but offset the vertical joints. Butter the ends of bricks before you lay them ❺. For each brick, note any squeeze-out and make adjustments for subsequent bricks to save time and mess. After you lay a few bricks, use a 4-ft. level to make sure your bricks are as level as possible ❻. Use a rubber mallet or the handle of your brick trowel to set the bricks ❼, or tap on the level to set them in place all at once.

Fold the wall ties into the mortar as you encounter them ❽. If there are no wall ties in the existing steps, install them every 16 in. After you set each course, take the edge of your trowel and scrape away excess mortar ❾. Throw the excess along the top for the next course. After you set a course, use a brick jointer to finish the joints ❿. If you need to add more mortar between the joints, use a pointing trowel and brick trowel and push the mortar into the joint. Let it dry for a bit before you use the jointer tool. If you are installing veneer on the sides of the steps, weave the bricks as you turn the corner ⓫.

Once you finish laying all the bricks, prepare for the stone treads by filling behind the bricks with mortar and smoothing the top. Scrape away excess mortar on top of the brick risers, and make sure all the joints look good; it will be difficult to correct them after the mortar dries.

As you complete each phase of the job, clean your work area. Remove all the brick and debris when you are finished to make room for the next phase.

>> >> >>

A cross section of the steps shows the brick risers and flagstone treads.

LAYING BRICK VENEER ON STEPS (CONTINUED)

1 Establish the brick-veneer location by marking 4 in. from the block face on both ends of each step.

2 Connect the marks with a chalkline to create a layout line for reference as you set the first row of bricks for each step.

5 Spread the right amount of mortar for each task. For each brick, note the squeeze-out and adjust accordingly to reduce mess.

6 Set several bricks and use a level to align them all at once.

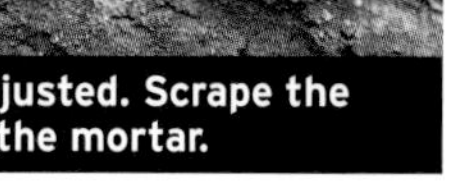

9 Scrape excess mortar each time a brick is adjusted. Scrape the trowel at an upward angle to avoid smearing the mortar.

10 Finish the joints with a brick jointer. You may have to wait for the mortar to dry slightly before you finish to avoid smearing it.

3 Establish a centerline for the stairs and mark it on each step.

4 Start each row at the centerline. For easier alignment, set the middle and end bricks and then lay the bricks between.

7 Make any adjustments by tapping on the bricks rather than pushing them into position.

8 Embed wall ties as you encounter them. Bend them into a layer of mortar, then spread more mortar to surround the tie on all sides.

11 Turn the corner by interweaving the long and short sides of the bricks.

TRADE SECRET

You can buy a special tool (called a brick hod) to carry bricks. But if you have to transport them only the few feet from the staging area to the work zone, press them together between your gloved hands and carry them horizontally. It's much easier to set them down this way.

STEPS WITH WALLS

BUILDING STONE STEPS COMBINES many of the techniques presented in this book into one project. In this way, it is one of the more challenging projects, but the rewards are potentially greater, as well. In this case, the steps, wing walls, and pedestal lanterns are a significant upgrade over the wooden steps they replaced. If you've successfully completed one or two of the other projects in this book, you're probably ready to tackle this one.

The project begins with removing the existing steps and covers installing an earth-formed footing, masonry-block steps, and stone veneer with treads. It includes how to build mortared wing walls in a dry-stack style and add electrical conduit to incorporate outdoor lighting. More about the techniques in this project can also be found in "Mailbox Column," p. 188, "Fire Pit with Seat Walls," p. 220, and "Stone Patio," p. 98.

GENERAL PREPARATION

SITE PREPARATION

INSTALLING CONCRETE BLOCK

PREPARING THE TREADS

SETTING STONE VENEER AND STAIR TREADS

CREATING WING WALLS

PRACTICAL CONSIDERATIONS

Building steps is not terribly complicated, but it does require a good deal of planning. For one thing, the entry that the steps serve will be unusable for the project's duration. If this is the house's main entrance, be sure to have the time, tools, and materials to complete the job without delay.

You'll also want to draw out a detailed plan. Think through the transitions at the top and bottom of the stairs, design details, the rise and run, and the amounts of materials you'll need to purchase.

Once you are familiar with the math principles involved in common stair building, the calculations for your steps will be straightforward. That said, building stairs does require a higher degree of accuracy than most masonry projects. Minor variations in height from step to step will be noticeable and may pose a tripping hazard. Check with your local building department to see if the staircase you are planning conforms to the building code. Even if your project does not require a building permit, it's important to build stairs that comply with safety codes.

For stairs attached to the house above a certain height, your local building code may require you to install handrails. If you are required to have handrails, it's a good idea to meet with a metal worker or handrail installer in the planning stage to figure out how the rail will meet the masonry steps. Some handrails are installed on top of the finished surface, such as brick or stone, while other handrail mounts are integrated within the substructure.

BEFORE YOU BEGIN

Planning all the details is essential for successful stair building. If you are removing existing stairs, make a note of the number of steps, stair height (rise), and tread depth (run). I like to make a sketch of the rise and run as well as the width and handrail location. Familiarize yourself with stair building (see "Do the Math" on the facing page), and draw an accurate plan that includes all the details.

As with any project, collect the construction materials in advance and establish an on-site staging area. For this project, you'll need room to mix concrete and mortar as well as an area to shape wall stones and stair treads. This staircase uses mortared dry-stack construction, meaning the stones are closely fitted together and the mortar doesn't show. This is a very popular style; however, it's not something you can throw together quickly. Plan on spending considerable time shaping stones for a tight fit. You'll also need a place to put stone debris. On this site, for example, there were drainage gullies that were a perfect place to deposit small stone shards and flakes.

And, as always, check with your local municipality and the site plan of your house, if you have one, for the location of underground water, sewage, electrical, and gas lines.

➔ **See "Establishing Staging Areas," p. 77.**

These steps may look like scrap wood, **destined to be replaced with stone, but they are actually a valuable resource. Before demolishing the existing steps, take careful measurements and save the dimensions; they will be very useful when building the new steps.**

Staging materials and tools **on the jobsite before you begin is the key to efficiency. To protect moisture-sensitive materials, such as bags of ready-mix concrete, stack them off the ground on pallets, cover them with plastic sheeting, and anchor the sheeting with heavy tools or block.**

DESIGNING STAIRS

When determining stair elevations on the plan, keep in mind that you will need to leave room for the mortar bed as well as the tread thickness. For example, if you are using 2-in. flagstone treads, you'll need to plan for at least 1 in. of mortar bed as well.

I like to extend the tread 1 in. past the riser. If you leave too much of an overhang, however, it creates a tripping hazard. Leaving just a little separates the tread from the riser, showing off a little more tread edge detail. This design is commonly used in more formal settings. In rustic settings, a single stone may serve as both the riser and tread. There is a lot less shaping involved with the latter.

This view clearly shows the relationship of the concrete block, mortar, risers, and treads. Note the thickness of the mortar and the treads' slight slope.

A 1-in. tread overhang, as compared to the more typical 3/4 in., creates a stronger shadow line under the nosing, which makes the stairs stand out visually. Do not, however, get carried away with deeper overhangs or you will create a tripping hazard.

DO THE MATH

Stair construction follows standards for comfort and safety. Typical step height (rise) is close to 7½ in. and tread depth is around 11½ in. (run). The height from step to step shouldn't vary more than ⅛ in. To determine how many steps you will need, divide the total rise by 7½ in., then spread the remainder evenly among the steps. To determine the tread depth, divide the total run by the number of steps.

For example, if the total rise from grade level to the top of the stairs is 36 in., divide that number by a comfortable riser height, say 7 in. The result is the number of risers (in this case, 5), plus a fraction. Divide the whole-number portion back into the total rise to get the exact rise per step (in this case, 7⅕ in.). You can always raise the grade at the bottom slightly if you need to make a slight adjustment. To determine the number of treads, measure the total run of the planned stairs. If the distance is 60 in., divide 60 by the number of steps (5), and you will need to make the tread depth 12 in.

I have seen a riser height as low as 5 in., and as high as 10 in., but these aren't comfortable steps. The smaller the riser, the more it presents a trip hazard. The same can be said of a tall riser. Most people are used to the standard of 7 in. to 8 in., so that is what they are naturally expecting when they walk up or down steps.

The same is true for treads. A short tread will not allow you to make a comfortable step with your foot. But if it's too long, you will have to take two walking steps before you make one rise, which is equally awkward. Typical treads are 10 in. to 14 in. People expect consistency for treads and risers.

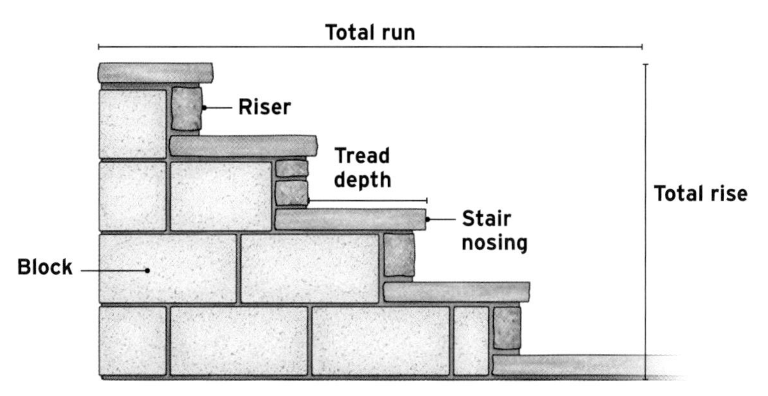

To measure total rise, determine the difference in elevation between the porch floor and the grade where the steps land. Divide that number by the desired height of each step, in inches, to calculate the number of steps.

DEMOLITION

Before laying concrete block, you must remove any existing stairs, excavate soil for an earth-formed footing, and pour the concrete footing.

When doing demolition work, there is always the question about what to salvage. Reusing materials often works to your advantage, first because it can save you money, and second because it is not much more work to set materials aside than to dump them. If your local refuse center charges by weight, rather than volume, a lighter load costs you less. In cases like these stairs, there was not much to salvage. While I don't like to throw out good materials, in strict terms of price of labor vs. cost of lumber, salvaged lumber is almost always more expensive than buying new.

If there is no reason to save the steps, use a long pry bar and sledgehammer to dismantle them ❶. Smaller tools will work, too, with more effort. Avoid using a wooden-handled hammer; all-metal hammers better withstand the abuse of demolition ❷. As you work, remove old nails and break off screw points to avoid puncture injuries. You may also need a socket or wrench set to remove lag screws and bolts.

On this house, the stone foundation façade ran behind the original staircase. You can, of course, lay the new stone up to the old, but I feel this type of joint is never quite successful. Even though it is more work, I prefer to remove enough stone from the old veneer to fully weave the inside corner. Even if the new stone doesn't match

1 Use a long pry bar and sledgehammer to remove wooden stairs. A reciprocating saw can come in handy, as well.

2 Pry between the treads with a metal-handled hammer if you have only small tools.

3 Wedge a chisel between veneer stones that need to be removed, and keep maneuvering them until they pop free.

4 When you have created access, wedge a chisel behind the veneer stones and pry them off. Cut wall ties before prying when possible.

perfectly, it is better not to have prominent joints where the stairs and foundation meet. Weaving makes the addition less obvious.

To remove dry-stack stonework like this, start at the top. Work a chisel into the joint until a stone pops free ❸. With properly installed stonework, every rock needs to be loosened. If the stone veneer covers concrete, you can sometimes work a chisel behind the stone and pry it free ❹. If the stone is hard to remove with a hammer and chisel, you might need to use an electric demolition hammer.

See "Demolition Tools and Equipment," p. 43.

Once you've removed the existing steps, stage materials close to the site. Cover finished surfaces with plywood and plastic to prevent damage. Cover shrubs with plastic, and tape finished surfaces to prevent cement stains. If you are working close to windows, cover them with foam board or plastic to prevent damage from airborne shards.

INSTALLING THE FOOTING

In this book, we show a mortar footing on gravel (see "Fire Pit with Seat Walls," p. 220), a wood-formed footing with rebar reinforcement (see "Block Wall with Stucco," p. 156), and an earth-formed footing (see "Brick Walkway," p. 174). The footing you choose for your stair project will largely depend on the existing conditions.

For the project shown here, we decided on a simple earth-formed footing with a layer of compacted gravel at the bottom and a thick layer of concrete with rebar for strength.

See "Construct a Basic Earth-Formed Footing," p. 140.

When installing a footing adjacent to a house foundation, it is a good idea to tie the two together with rebar. This will help prevent cracks in the masonry joints if settling occurs. To do this with $\frac{1}{2}$-in. rebar, bore $\frac{5}{8}$-in.-dia. by 6-in.-deep holes every 16 in. into the existing foundation. Position the holes so they are located in the middle of the step footing. Use epoxy to fasten the rebar to the foundation. It should extend well into the footing's rebar grid. Tie the rebar grid together with wire.

For small footings that require less than a yard of concrete, I usually opt to mix the concrete by hand in a wheelbarrow or motorized mixer (see p. 62). However, for footing pours of over a yard, I generally call a concrete truck for convenience. Ready-mix concrete is usually sold by the yard and is much more cost-effective when sold in large quantities; having it delivered is much less taxing on your body. There is usually a small delivery fee. >> >> >>

Connect the new footing to the existing foundation by boring holes into the old foundation, inserting rebar, and securing the rebar with epoxy. Tie the rebar grid together with wire.

CONSTRUCT A BASIC EARTH-FORMED FOOTING

A simple earth-formed footing is sufficient for many masonry projects, and it's easy to build, as well. To begin, mark the footing perimeter 6 in. beyond the face of the project on all open sides. For example, if your project is 5 ft. wide and extends 5 ft. from the house, your footing will be 6 ft. wide and extend 5½ ft. from the house. Dig the hole at least 16 in. deep (or as recommended by code for your region) and carefully trim the sides so they are vertical and square and the base is flat. The top of the footing needs to be slightly below finish grade so you can backfill over the foundation, leaving a nice, clean surface with mulch or grass.

Thoroughly compact the base soil with a tamper. Add a 4-in. layer of gravel and compact it thoroughly. Pour in an even 4-in. layer of well-mixed concrete. Lay ½-in. rebar in an 8-in. grid on top of the concrete, tying the rebar with wire at the intersections. Add another 4-in. layer of concrete, and smooth and level the top so it's about 1 in. below grade. Let the concrete cure for 24 hours before laying stone, block, or brick on top.

❶ **Mark the perimeter of the footing 6 in. outside the project footprint.**

❹ **Compact the added layer of gravel.**

❺ **Pour in a layer of concrete over the gravel.**

❷ Dig the footing hole to the depth required by code (12 in. in our area).

❸ Compact the soil firmly; a concrete block works fine as a tamper.

❻ Add a layer of rebar and secure it with wire at the intersections.

❼ Add a second layer of concrete and smooth the top surface.

MAKING A MOCK-UP

Concrete block serves as the supporting structure for the steps, but you won't see it once it's covered by stone. As part of your detailed plan (see p. 136), you will want to calculate the thickness of the treads and the stone facing, as well as the mortar beds. Measure from the finished dimensions to establish the sizes of the block tops and faces. Doing this on paper can be tricky enough, but transferring it all to the real world can be a definite head-scratcher. Typical block sizes (4 in., 6 in., and 8 in.) are not standard step heights. You might have to double up on blocks, cut some short, or get creative to make this work. There is no set formula. To ease the process, I always create a mock-up before I spread mortar. In addition, I like to have blocks of various sizes on hand, so I can mix and match as necessary ❶.

➔ **For more on concrete block, see "Block," pp. 58–60.**

When making a mock-up, allow ½ in. to ¾ in. for the mortar joint between the blocks when measuring ❷. You may have to elevate the bottom block more than a typical joint size, but you can always make up for this with the mortar bed.

Once the block height is established for each step, place the blocks to attain the proper tread depth ❸. Remember to account for the stair nosing when calculating tread depth. Spend as much time working these blocks into place as needed until you are satisfied and have double-checked the measurements with the plan. Before taking down the mock-up, note the block configuration on your plan and mark the block locations clearly on the footing surface. Even though this process may be time-consuming, it will save you unnecessary headaches down the road.

USING A STORY POLE

If you've installed the footing surface level, creating a story pole can make installing the blocks to the correct elevation much easier and more accurate. A story pole shows all the critical elevations marked on one board. It allows you to check your accuracy as you build without repeatedly measuring with a tape, which can introduce errors.

1 Assemble a mock-up of different-sized blocks to establish rise and run measurements.

2 Pay special attention to measurements at the top and bottom of the run.

3 Position the concrete blocks to achieve the proper tread depth.

LAYING BLOCK

Each row is essentially a separate block wall, or "step wall," if you will. If you haven't already read the "Block Wall with Stucco" chapter (p. 156), it's a good idea to do so before continuing this chapter. Also, read the section on mixing mortar and setting block.

When building any block wall, it is important to level the first row in both directions. If the footing surface has unwanted pitch or irregularities, now is the time to correct it by setting the first course level.

Begin by setting the row's two end blocks, aligned with the layout marks. Then set the blocks between them using a level or string to guide alignment ❶. Unless you have planned it perfectly, you will need to cut a block to complete the row. Set it in place with plenty of mortar and align it using a level ❷.

➔ **See "Cutting Concrete Block," p. 60.**

Turn the blocks on the ends perpendicular to the horizontal step rows to begin the sidewalls ❸. The sidewall will help tie individual step walls together and make the whole assembly stronger. Start the second course by weaving the sidewall into the first step wall ❹. Complete as much of the tallest step wall, along with the sidewalls, before building the next step wall. This minimizes the amount of reaching or stepping over the work you have to do as you lay more block ❺. In the joint under the top row of bricks, insert wall ties every 16 in. Also insert wall ties in the outside sidewall joints ❻.

For this project, measurements allowed us to top each step wall with a 4-in. cap block ❼. If you can't use a cap block, fill the block cells with mortar and rubble. Also, fill the gaps between the walls with rubble; then pack mortar on top of it ❽. Smooth the mortar with a trowel so the top of each step is flat ❾. >> >> >>

1 Set the end blocks of the first row. Then set the filler blocks, using a long level to align them.

2 Cut a spacer block, butter it with plenty of mortar, and set it to complete the row.

When building a block wall, **always level the first row in both directions.**

3 Turn the corner at the end of the first step wall to begin the sidewall.

LAYING BLOCK (CONTINUED)

4 Start the second course by alternating the direction of the corner blocks. This ties the walls together and strengthens the assembly.

5 Lay the second row in the same manner as the first; begin at the ends, lay middle blocks, and align them with a level.

6 Insert wall ties in the mortar joints every 16 in. where you plan to lay veneer.

7 Cap each step wall with 4-in. blocks. If your measurements don't allow for cap blocks, fill the block cells with rubble and mortar.

8 Fill the voids behind the step walls with rubble, excess soil, or broken pieces of concrete block.

9 Pack mortar into the void with a trowel; smooth the mortar so it's flush with the cap blocks.

SHAPING STONES FOR TREADS

You can buy precut stone treads for stairs or shape your own. Buying precut treads saves considerable time, but the treads are more expensive and need to be ordered with the exact dimensions well in advance. If the precut treads are too large, you can cut them with a saw, but if they are too small, you will have to reorder them. We chose to cut our own treads from some flagstone that was relatively square and had many straight edges to work with.

For this project, the tread depth, or the distance from the front of a tread to the face of the next riser, was 12 in., including the overhang. A joint directly below the riser calls attention to itself, so we selected stones we could shape to be either 10 in. from the nosing (the long front edge) to the back edge, or 13 in.

To shape flagstones into treads, you'll need a straightedge and a brick hammer. Begin by selecting a stone with a long front edge that also has the desired width. Draw a straight line along this edge that minimizes waste ❶. With a brick hammer, shape the stone's edge close to the line. If you need to remove a lot of material in the first pass, stay well away from the cut line ❷ because the stone may break in unexpected ways. After the first pass, or when the edge is even but the flakes are becoming a little harder to remove, turn the stone over. Draw a straight line opposite the first and repeat the process ❸. Avoid removing too much stone at once to reduce the risk of an errant break. You may have to flip the stone several times to approach the line with control. Stand the stone upright to work the ends ❹. This reduces the chance of an uncontrolled break across a stone's face.

Once you're satisfied with the front edge, use a framing square to mark the sides and back of the tread, then cut the tread square with a grinder or circular saw fitted with a masonry blade ❺. Make sure you're wearing gloves, goggles, and hearing protectors for this cut.

To incorporate a chase for electrical lines into the block step walls, see "Install a Chase," p. 154.

Stack all the treads close to the work area, and try to limit the number of times you move them ❻. Breaking a tread can be costly and severely disappointing. It's a good idea to cut all the treads ahead of time before moving on to the next step.

>> >> >>

1 **Draw a straight line across the stone's longest edge in a way that makes the best use of the stone and minimizes waste.**

2 **Remove stone to the line with a brick hammer. Removing small amounts of stone at a time helps control the breaks.**

SHAPING STONES FOR TREADS (CONTINUED)

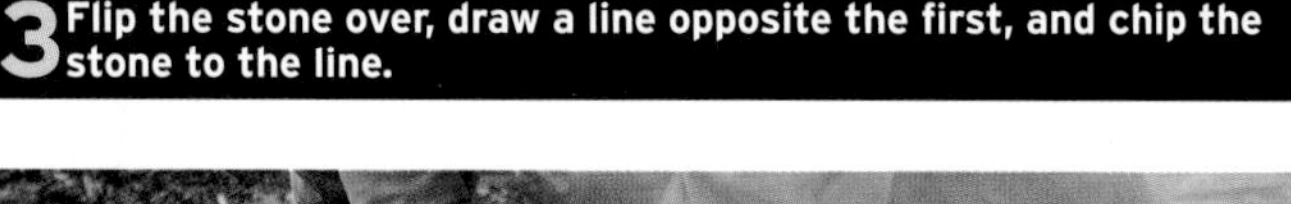

3 Flip the stone over, draw a line opposite the first, and chip the stone to the line.

4 Remove waste stone from the end by standing the stone on its opposite end and striking downward, toward the center.

5 Cut the stone's sides and back profiles with a circular saw fitted with a masonry blade.

6 Stack stones near the work zone but out of the way. To avoid damage, try to limit handling the stones unnecessarily.

TRADE SECRET

Mark the tread's elevation on the block riser before setting treads. This provides a reference when setting and aligning tread stones.

SETTING RISERS

Dry-stack veneer looks better with a minimal joint (a joint without visible mortar) between the riser and tread. To accomplish this, start at the bottom of the stairs and work your way up. This allows you to set the risers on the treads, rather than behind them.

➔ **See "Laying Brick Veneer on Steps," p. 131, to learn about an alternative method where risers are set behind the tread.**

The mortar in dry-stack stonework is hidden deep inside the joints; stones should fit closely together to imitate traditional dry-stack, unmortared stonework.

1 Spread mortar for the riser stones at each end of the first row. Hold back the mortar from the riser face location.

2 Set the riser stones 6 in. from the block face, or at whatever distance fits your stair design.

3 Level across the block's top and leave room atop the riser stones for a 1-in. layer of mortar for the tread.

We chose to establish our riser face 6 in. in front of the concrete block's face, because some of the stones were rather bulky and we needed the extra space. However, if the stones you are working with are smaller or flatter, you can opt for a 4-in. or 2-in. veneer thickness. Veneer thickness will affect the transitions at the top and bottom of the stairs, too, so make sure you plan for this ahead of time.

Before beginning the risers, shape the majority of the veneer pieces you'll need into rectangular stones that will fit together snugly. With the stones shaped, you're ready to mix the bedding mortar.

See "Mortar Recipes," p. 61.

Begin setting riser stones at the ends of the bottom row. Spread a layer of mortar 1/4 in. to 1/2 in. thick for the first riser stones ❶. Using a tape measure for reference, set the riser stone with its face exactly where you need the riser to be (in our case, it was 6 in.) ❷.

>> >> >>

SETTING RISERS (CONTINUED)

For taller stones, check that there is enough space to allow a 1-in. mortar joint under the tread ❸.

Once the ends of the first row are set, set a riser stone at the center; use a long level to align it with the end stones ❹. Fill the remaining space, again using a level for alignment ❺. If you find that a stone is too high, take time to trim it with a brick hammer. One stone set too high can prevent the tread from being set correctly.

As you progress, tap each stone so it sets fully into the mortar bed. Fill behind the stone with mortar, pack mortar into crevices, and smooth the top with a trowel ❻. Embed ties in the mortar as you encounter them ❼. When the mortar has dried slightly, go back and scrape away any mortar that has squeezed out between the stones and would be visible later.

WHAT CAN GO WRONG

Mortar stains are tough to deal with but easy to avoid. When setting tread stones, use a drier mix of mortar and hold it well back from the edge when spreading it.

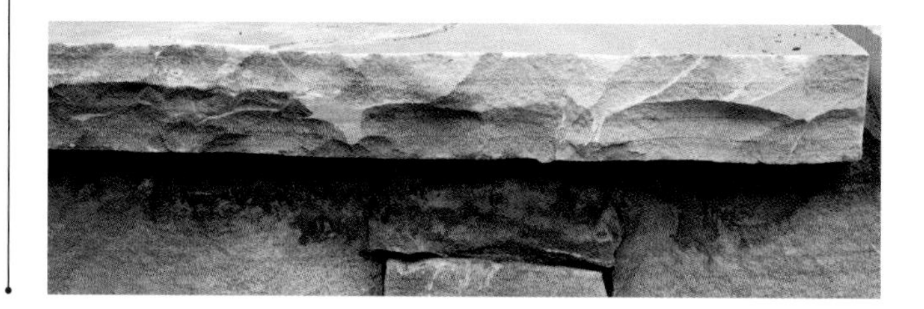

4 Set a riser stone in the center of the first row, and level across to be sure it's set to the same height as the end stones.

5 Lay the rest of the riser stones. Frequently check for level, and be sure none of the stones will interfere with setting the tread.

6 Fill behind the riser stones with mortar, push mortar into all the crevices, and smooth the top flush.

7 Pack mortar around wall ties as you encounter them. Bend the ties and embed them close to the stones but not against them.

SETTING STAIR TREADS

1 Align the tread with the reference line and set with a mallet.

2 Use a level to determine the pitch of the tread.

The process for setting stair treads is essentially the same as for setting patio edges (see "Setting Edge Stones," p. 107). However, to set treads successfully, you need to be very accurate and develop an uncompromising standard. If, for example, a tread is set ⅛ in. too low, it must be reset. Or if you find a riser stone that's too high and interferes with how a tread sets, remove, trim, and reset it correctly.

At one end of the step, spread an even layer of mortar that's enough for one stone, and tap a tread stone into place with a rubber mallet ❶. Level the stone's leading edge but create a ½-in. slope from back to front ❷. This prevents water from pooling on the steps and is more comfortable and ergonomic to walk on than a level tread. Set the overhang at 1 in. from the riser face ❸. Once the tread stones at each end are set, use a level or stringline to set the tread stones between them ❹. Again, take extra care to align the top surfaces and leading edges. Any variation creates a tripping hazard.

Scrape mortar from between the joints. Then allow the treads to set for 24 hours before grouting the joints. At this point, it is a good idea to cover the stairs to prevent anyone from walking on them before the mortar has a chance to fully set. A simple tarp held in place by stones will do the job. The tarp also protects the mortar in case of an unexpected rain shower ❺.

3 Overhang treads 1 in. over the face of the riser.

4 Fill in the treads between the ends, using a level to keep everything aligned.

5 Cover the stairs with a tarp when finished.

DESIGNING MATCHING WING WALLS

Adding curved wing walls to stairs significantly enhances their appearance. Technically, wing walls are no more difficult to construct than the walls in the "Fire Pit with Seat Walls" chapter (p. 220). If you haven't read that chapter, it's a good place to start before tackling any mortared wall project.

There are, however, some key differences between wing walls for stairs and a stand-alone mortared wall. Most importantly, wing walls for stairs must match each other. They must also be joined to the house and woven into the stairs. We designed the wing walls shown here with a sloped cap that runs from the bottom of the stairs to the porch at top and incorporated outdoor lighting, which is detailed on pp. 153-154.

There is no formula for designing wing walls. They can be level, step down, or slope down, as shown here. They can be straight, or curved, or just about anything you want them to be. Whichever design you choose, it's a good idea to sketch ideas out on paper before you begin laying stone. The project shown here places the height of the wall at the top of the steps level with the porch. Then the wall slopes down the steps with a slight curve and levels out at the bottom.

To begin laying out the wall, set veneer stones on one side of the stair treads to represent the wing wall's curve ❶. Adjust the stones until you are satisfied with the curve. From the inside sidewall face that you just established, measure a 12-in.-thick wall and mark the outside sidewall face on the footing with marking spray ❷. Use the centerline (measuring back from the nosings) to mark the wall's mirror layout on the other side.

CREATING A MIRROR IMAGE

To help create matching walls, establish reference points based off a centerline at easy-to-measure locations, such as stair nosings. Once the inside wall face is established, measure out 12 in. and mark the outside wall face every 12 in. or so on the footing. It is not critical that the outside curves match exactly because they can never be seen from the same viewpoint.

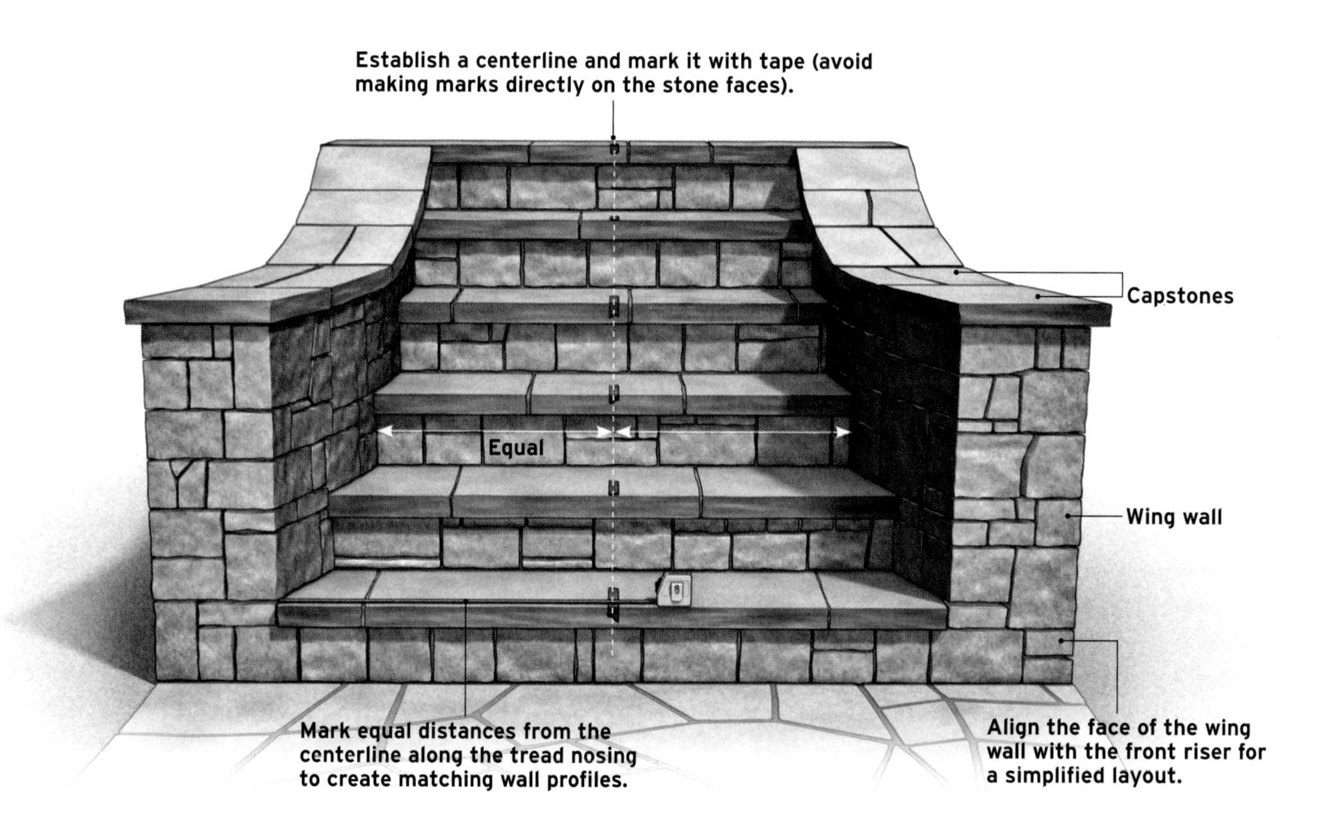

BUILDING WING WALLS

Start building the wing walls by setting perimeter stones in a mortar bed. If you are incorporating lighting in the walls, you will need to install conduit where you want the light (see p. 153). If the conduit exits the wall at grade level, leave a small gap between the stones and align the stone faces with a level ❸. Continue building the wall by setting the corner stones first, then shaping stones to fill gaps ❹. With dry-stack style, try to keep the mortar back 1 in. from the stone faces while you set stone. When setting the corner stones, it is not practical to use stringlines to align them, as we explain for the mailbox column (see p. 188). Simply use a level to align the edges and set the corner stones plumb ❺.

To integrate sidewalls and stairs, weave the wall stones to fit the stairs' profile ❻. Take time to fit stones closely to the risers and around the tread nosings for best appearance and function.

The wing-wall design calls for the underside of the capstones to end flush with the underside of the top tread. Use a stringline, fixed at the cap's endpoint, to mark the trim line where the wall stones meet the cap ❼. Where the wing wall meets the house, the wing wall's veneer stones should interlock with the house foundation veneer ❽. This makes for a far more natural-looking joint. It also reinforces the wing wall and reduces the chance of cracks or failure. Before the mortar sets, remove visible excess from between the joints with a pointing trowel ❾. >> >> >>

1 Position wall stones on the stair treads to establish the wing wall's location and contours.

2 Measure for a 12-in.-thick wall, and mark the outside wall faces on the concrete footing for both the left and right wall.

3 Lay the perimeter stones at the wall base. Leave a gap between the stones if you're running conduit for lighting or receptacles.

4 Position wall stones on the stair treads to establish the wall's location and contour.

BUILDING WING WALLS (CONTINUED)

5 Plumb the corner-stone edges with a level. Be precise, because any variation will draw attention.

6 Weave the wall over the stairs. For best results, carefully shape wall stones to closely follow the tread and riser profiles.

7 Use a stringline to locate the underside of the capstones. Mark a cut line on the dry-fitted wall stones, and trim them carefully.

8 Weave the wing wall (left) into the house veneer (right), alternating the stones to lock everything in place.

9 Scrape out excess mortar from the joints before the mortar sets fully.

ADDING ELECTRICAL CONDUIT

Adding electrical conduit for exterior lighting or for a receptacle is not difficult. Even if you are not planning to do the electrical work yourself, you will need to install the conduit. For this project, you'll need 4-ft. and 2-ft. lengths of 1-in.-dia. electrical conduit fitted to a 90-degree elbow.

➔ **See "Install a Chase," p. 154.**

The aim is to embed the conduit in the wing wall so that an electrician can run wire through it and connect it to a light mounted on the column. To do this, hold the conduit roughly centered on the footprint of the wing wall ❶. At the bottom, sink the conduit deep enough in the footing mortar so the end will exit the wall at or below grade ❷. To secure the conduit while you construct the wall, tie it to one of the treads ❸.

Keep the conduit centered as you build the end column. Typical conduit is not very flexible, so it may interfere with setting stones at the top if you don't keep it centered along the way ❹. When setting the capstones, plan to have a joint where the conduit emerges from the column; this is much easier than having to drill a hole in the middle of the stone. Notch the stones around the conduit ❺, and then grout the joint ❻. Scrape the mortar around the conduit flush to the capstone so the light fixture base will seat properly. >> >> >>

TRADE SECRET

Seal the conduit ends with duct tape to prevent dirt and mortar from clogging the conduit. Your electrician will thank you for this consideration.

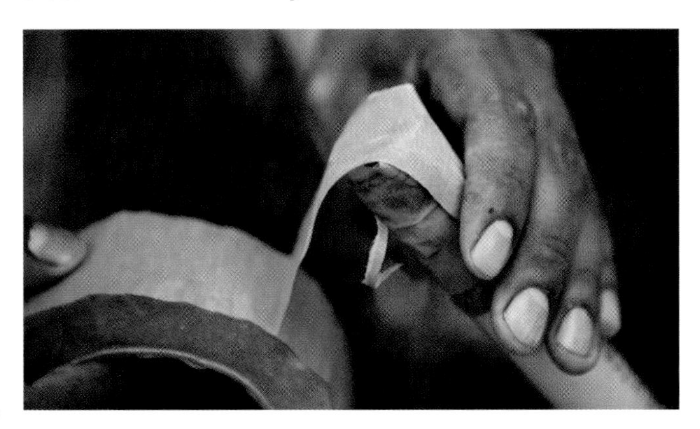

1 Position electrical conduit in the center of the wing wall. We used two lengths of $1\frac{1}{2}$-in. conduit fitted to a 90-degree elbow.

2 Set the conduit elbow low enough so the end exits at or below grade.

ADDING ELECTRICAL CONDUIT (CONTINUED)

3 Temporarily tie the conduit to a stair tread to hold it in place while you work.

4 Build the wall around the conduit. Frequently check the conduit to be sure it remains centered.

5 Orient capstones so the conduit will fall in a joint. Then notch the stones to accommodate the conduit.

6 Grout around the capstones, and scrape the grout flush so that the light fixture base will rest on a level base.

INSTALL A CHASE

No matter what your current landscaping needs are, it is a good idea to install a PVC chase under the steps through which you can pass electrical and irrigation lines later. Simply tape the ends of a 2-in.-dia. PVC pipe, and embed it in the mortar under the steps. Use a length of pipe that extends well beyond the stair sidewalls; you can easily cut it back later. However, make sure the pipe is located below finished grade so it's not an eyesore.

FINISHING THE WALL WITH CAPSTONES

When you have finished building the wing walls, make sure they are smooth on top and level from side to side. The remaining work of cutting and setting the caps will be just a matter of marking the angles with a pencil and cutting the stones with a saw or hammer.

Span as many of the capstones across the wall as possible. Spanning the wall with one stone is more difficult than using two stones with a joint in the middle because the opposite edges need to be shaped perfectly and suitable stones are harder to find. However, even if you can span only every third stone, it will look better than having a continuous joint down the middle ❶.

Dry-fit stones on top of the wall and mark cut lines with a pencil. In this case, the 12-in.-wide wall has a 14-in.-wide cap with a 1-in.-deep overhang on each side. Use a grinder or circular saw to cut the stone 15 in. wide ❷. That will leave you ½ in. per side to work the edges of the stone with a hammer or chisel ❸.

After setting the caps in mortar and allowing them to dry, grout the joints just as you did the treads. Use your brick trowel and pointing trowel to push the grout into the joints. Let the grout dry for a while before scraping it back with the blade of your pointing trowel ❹. Let the mortar cure for 4 or 5 days before mounting the lights.

➔ **See "Dry-Mix Grouting," p. 114.**

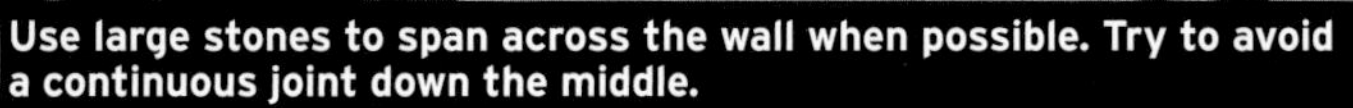

1 Use large stones to span across the wall when possible. Try to avoid a continuous joint down the middle.

2 Cut the capstone 1 in. wider than its finished width, so the edges can be textured with a brick hammer.

3 Texture the capstone edges to remove most of the saw marks and any irregularities left on the edges.

4 Grout the capstone joints, including the joint between the wall and the capstones.

BLOCK WALL WITH STUCCO

A BLOCK WALL, FINISHED WITH stucco, serves the same purpose as a brick or mortared stone wall, but it is much easier and more affordable to build. Masonry block, sometimes called cinder block or concrete block, is readily available and easy to transport. Once you start building the wall, the blocks lay in place quickly. With a little patience, most people can achieve professional-looking results on their first project. The stucco finish is perfectly suited for small projects, too. Stucco can be mixed in small batches and, after first applying a concrete brown coat, can be troweled directly to the block wall without lath or building membranes. To visually tie a wall to other stonework on your property, you can cap it with similar stone.

Given their proximity to parking pads, some walls need beefing up with rebar to help prevent cracking should they get bumped by a vehicle.

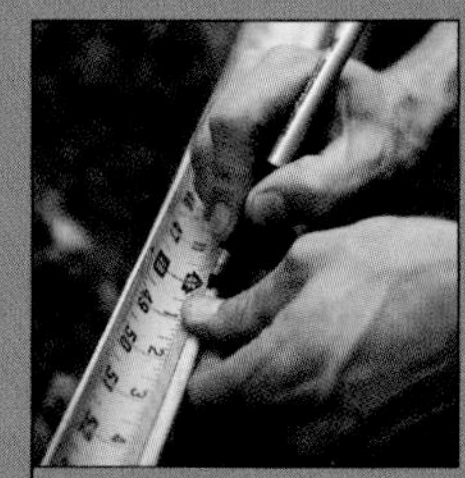

PRACTICAL CONSIDERATIONS

It's never safe to assume that you won't hit an electrical wire or water line when digging, even if you're digging only a few inches down. The presence of an outdoor light, for instance, is a pretty good indication that you might encounter an electrical wire. Carefully dig out the footing by hand instead of using a backhoe. You never know what else you might find—such as a water pipe at footing depth. Before covering anything, note all utility line locations on the property survey map.

The end of a wall is a natural location for outdoor lighting. If you're going to add an outdoor light or think you might do so in the future, see "Adding Electrical Conduit," p. 153.

Stucco is a permanent finish, but it's easy to apply a stone or brick veneer over it in the future should you want to make changes. If you think you might add a veneer later, remember that the veneer thickness will change the cap overhang, and you may have to re-lay the cap—or lay a new cap—to achieve an acceptable overhang.

See "Brick Veneer," p. 118.

A light on an existing patio (as here) is a good clue that you might find an electrical wire; sure enough, there was a wire no more than 2 in. below the surface.

Water pipes are among the things you might find when you excavate. You can deal with discoveries like this during the project or later on. Mark the pipe's location on the property survey map for future reference.

BEFORE YOU BEGIN

One advantage to building with block is that it is easy to envision the finished height of the wall. While planning your structure, you can loosely stack blocks and place a capstone on top, adjusting the height of the wall as needed until you're happy with the look.

Before excavating, it's a good idea to decide how to lay out the project. In this book, we describe a couple of different ways to do this, including using a garden hose.

➔ **See "Measuring and Marking Tools," p. 32.**

For some projects, it's better to use batter boards. Batter boards enable you to locate the exact position of the wall using stringlines (mason twine) for reference. You can remove the twine during excavation and replace it in the same location later when it's time to position form boards and wall blocks. There's no need to repeatedly pull measurements.

The last thing to do before you begin is to check the weather. This is not so important for the day you lay block; a little rain is not a big deal. However, on the days you apply the cement brown coat and stucco finish, be sure there won't be any rain. Even a light sprinkle can destroy a stucco finish.

WHAT YOU'LL NEED

- 24 regular 8-in. blocks
- 10 bags concrete mix
- 1/2 cu. yd. masonry sand
- 1 bag portland cement for setting caps
- 2 bags portland cement for stucco
- 2 bags Type S masonry cement for mortaring block
- 1/2-in. rebar for footing
- Rebar chairs
- Rebar tie wire
- 10 to 12 pieces stone for cap
- 15-gal. bucket of premixed stucco
- Bonding agent for stucco
- 2 10-ft. 2x6 (boards for forms)
- 4 2-ft. 2x2s (stakes for batter boards)
- 2 1x4s (crosspieces for batter boards)
- 3-in. screws
- Masonry tools
- Woodcutting saw
- Square plastic trowel for stucco application
- Square foam trowel for stucco finish

LAYOUT AND EXCAVATION

To set up batter boards, roughly locate one edge of the wall with a stringline ❶ and drive the batter-board stakes so the string intersects the crosspiece approximately in the middle ❷. Make sure you locate the batter board outside the area where you will dig the foundation. For the wall shown here, the sidewalk determined the wall location at one end, which made it fairly easy to locate the batter board. For larger projects or projects with more variables, locating the batter board may require some trial and error. Take the time to level the stringline using a line level ❸. A level stringline allows you to quickly check elevations by measuring down from the twine.

Setting the form

Build the footing form with 2x6 lumber; it doesn't have to be treated lumber, but it should be straight ❹. With a wall 8 in. wide and 8 ft. long (as shown here), you can build a form with two 2x10s. The inside dimensions of the form are 12 in. wide by 8 ft. long. » » »

1 The stringline across the batter board roughly locates the outside face of the wall.

LAYOUT AND EXCAVATION (CONTINUED)

To lay out the form perpendicular to the house, we used the 3-4-5 method. It's a great trick for setting one line square (90 degrees) to another. First, make a mark on the ground 3 ft. from the corner of the form ❺. Then make a mark on the form 4 ft. from the corner ❻. Run a tape from one mark to the other and move the outboard end of the form until the distance between the marks is exactly 5 ft. ❼.

When you get the form in the right position, make marks under the stringline at both ends of the form ❽. Use a level or plumb bob to ensure correct placement. Also mark the stringline location on the batter board ❾. This enables you to place the form in the exact same location after you dig. Before removing the form, mark the footing perimeter as well; it will show where to dig ❿.

Dig to the footing depth recommended by your local building department. In our case, 10 in. was deep enough to accommodate 3 in. of gravel, 5½ in. of concrete, and a couple inches space below grade. Compact the soil at the bottom of the footing excavation ⓫. After spreading 3 in. of gravel over the soil, compact the gravel too. For a small project like this, a hand tamper works fine ⓬.

➔ **For more on compacting tools, see p. 45.**

2 The stringline should intersect the crosspiece in the middle.

5 To ensure that the footing is square to the house, make a mark 3 ft. from the corner.

8 Mark plumb from the stringline on the top of the form at both ends. These marks will help you reposition the form later.

10 Mark the outside perimeter of the form with marking spray to show where to dig, and remove the form.

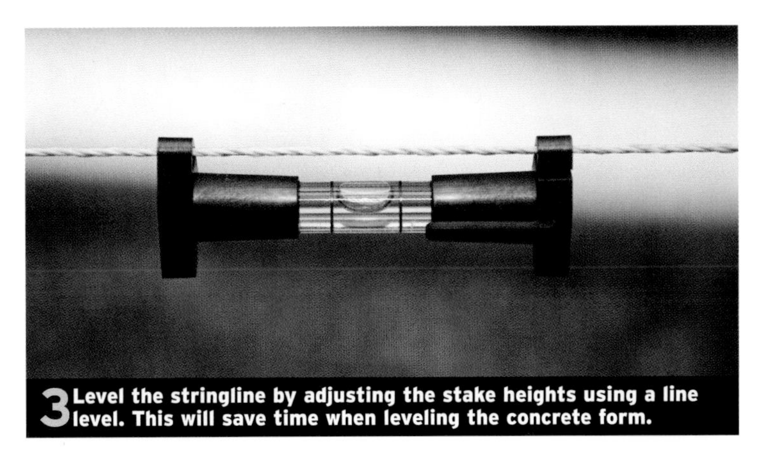

3 Level the stringline by adjusting the stake heights using a line level. This will save time when leveling the concrete form.

4 Build the form with 2x6s. For an 8-in. block wall, plans called for a 12-in.-wide footing.

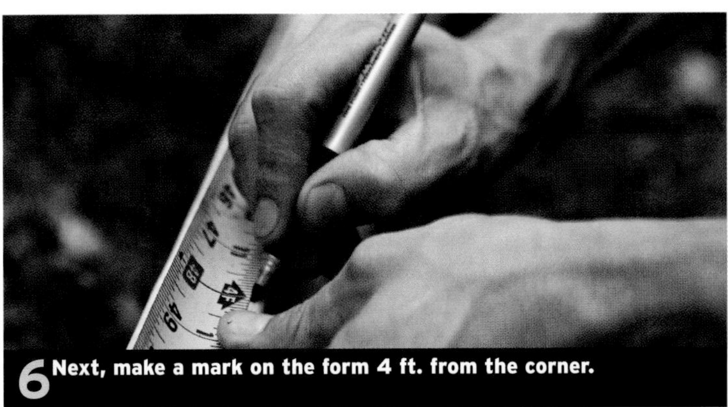

6 Next, make a mark on the form 4 ft. from the corner.

7 Have a helper shift the outboard end of the form until there is exactly 5 ft. between the two marks.

9 Mark the crosspiece of the batter board where the string intersects it.

11 Dig the footing trench to the required depth and a couple of inches wider than the form. Then compact the soil at the bottom.

12 Evenly spread gravel in the trench to a depth of 3 in., and compact it with the hand tamper.

POURING A CONCRETE FOOTING

When you're ready to put the footing form back in the trench, replace the stringlines and align them with the marks on the batter boards. Place the form in the bottom of the trench. Then use a level or plumb bob to align the marks on the form with the stringline; the marks should fall directly under the stringline ❶. Before moving on, pile a little soil around the outside edges of the form to keep it in place. Keep the top of the form level by bracing it with a little gravel underneath, if needed ❷. If you leveled your stringline earlier, you can adjust the form in the long direction by measuring down from the stringline ❸.

1 Place the footing form in the trench and re-establish the stringline. Then, use a plumb bob or level to position the footing form.

Setting the rebar

Adding rebar to a small wall like this may seem like overkill, but rebar doesn't cost much, doesn't take much time to install, and is well worth it for the added strength it delivers. To keep the rebar in the middle of the pour where it adds the most strength, support it with rebar chairs spaced every couple of feet ❹. Lay the long ½-in.-dia. rebar pieces on the chairs, leaving at least 1-in. clearance between the end of the rebar and the form ❺. Every 8 in., tie 10-in. rebar crosspieces to make a rebar ladder ❻. Keep the rebar 1 in. away from the form. For this wall, we cut the uprights to 16 in. and tied them to the rebar ladder every 16 in. ❼.

For more on rebar, see p. 68.

4 Place chairs for rebar every couple of feet in the form and within 5 in. of each end.

Adding the concrete

Filling the form shown here requires 10 bags of concrete. That would be quite a bit to mix by hand, so it's a good idea to rent a small electric mixer. When using a mixer, add a small amount of water to the mixer barrel before you pour in the concrete. This helps keep concrete dust from sticking to the sides. For a small mixer, mix one bag at a time to avoid overwhelming the motor. When adding concrete, put the open end of the bag entirely in the mouth of the barrel and

» » »

TRADE SECRET

A quick way to measure rebar is to use the form instead of a tape measure. Place the rebar with one end butted against the inside of the form. Use a grinder with a metal-cutting blade to cut the rebar 4 in. short of the other end. This leaves a 2-in. clearance between the rebar and the form at both ends.

2 Level the footing form in both directions; push gravel under the form to make the necessary adjustments.

3 Measure down from the stringline at both ends of the footing form to check that it is still level.

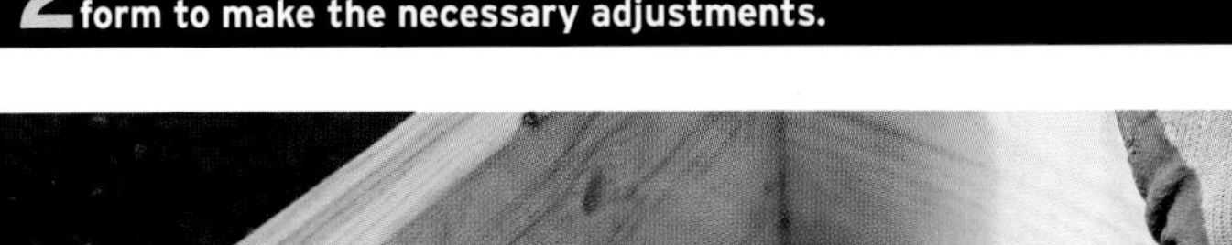

5 Place two ½-in.-dia. rebar rods on the chairs. Check to be sure there is at least a 1-in. clearance between the rod ends and the form.

6 Use rebar ties to fasten the 10-in.-long crosspieces to the longer sections of rebar every 8 in.

7 Tie 16-in. vertical sections of rebar to the rebar ladder every 16 in. Avoid burying the rebar ends into the soil below the gravel.

8 Mix enough concrete to completely fill the form. To keep the mix consistent, carefully measure rather than estimate the water.

POURING A CONCRETE FOOTING (CONTINUED)

9 Settle the concrete in the form by repeatedly plunging a length of rebar into the wet concrete.

pour slowly to keep the harmful dust to a minimum ❽. Take precautions, such as wearing a dust mask. Add small amounts of water until the concrete is thoroughly mixed and is the consistency of stiff cottage cheese.

When the footing form is full, settle the concrete by repeatedly plunging a length of rebar into the mix ❾. This will help remove big air pockets. To settle the concrete further, vibrate the form by tapping on its side or by mechanically vibrating the form with a power tool ❿.

After the concrete is settled, screed the top with a straight board to remove excess concrete. The top does not have to be perfectly smooth, but it does have to be flat. Let the concrete set overnight before laying the block ⓫.

10 Vibrate the footing form with a power tool, such as a circular saw or an impact drill, to fully settle the concrete.

11 Screed the top of the concrete flush with the footing form.

TRADE SECRET

It takes the same amount of time (roughly 2½ minutes) to mix an 80-lb. bag of concrete in a wheelbarrow as it does in a mixer. Considering the concrete in a mixer must be transferred to a wheelbarrow to get it to the form, using a mixer will not save you any time—but it may save your back.

LAYING THE FIRST COURSE OF BLOCK

Laying block is not difficult. However, it is important that each block is straight and plumb. Any irregularities will show through the stucco finish.

If you removed the stringline while pouring the footing, replace it now. Check that the form did not move during the pour. Then use the stringline to locate the position of the first block ❶. If the string represents the finished face of the wall, set the blocks back 1 in. from it to allow for the thickness of the brown coat and finish stucco coat. Position blocks at both ends. With a straightedge, mark the face of the wall where it will meet the footing ❷. If the concrete is still soft, you may be able to scratch the line into the surface.

Mix a batch of mortar by hand or in a mortar mixer consisting of 1 bag of Type S masonry cement and 18 shovels of sand. If this your first time laying block, start with half a batch of mortar.

Using your line as a reference, spread two 1-in.-thick rows of mortar down the footing inside the wall's perimeter ❸. You can set a block next to the mortar to help you see where to spread mortar (under the center rib and ends of the block) ❹. Once the mortar is spread, hold the block by the center, gently lower it over the rebar, and align it with the reference line ❺.

When working with block, be gentle. It is easy to push too hard on the block and squeeze out all the mortar. (If this happens, remove the block and replace the mortar.) To avoid squeeze-out, adjust the blocks by tapping gently with a trowel ❻. Be careful not to over-adjust. Every time you tap the block, a little mortar is squeezed out.

Before setting the next block, butter one end ❼. Lower the block at an angle so it meets the previous block and the mortar bed at the same time ❽. After tapping the block into the mortar, use a mason level to align the blocks along the top and face ❾.

At the end of the row, cut the last block to fit. If you don't have a cutting tool (see "Cutting Tools," p. 38), it's easy to accurately break the block with a brick hammer. Tap repeatedly along the cutline all the way around the block until you've made a shallow groove ❿. One or two blows should be enough to break the block along the groove ⓫. Butter the rough end and orient it to the inside when you set it ⓬.

>> >> >>

1 Re-establish the stringline and then measure back 1 in. to locate the block face.

2 Temporarily position the end blocks and score a line on the footing to indicate the position of the wall.

LAYING THE FIRST COURSE OF BLOCK (CONTINUED)

3 Spread mortar inside the wall perimeter—but only enough for the first two or three blocks.

4 Rest a block just outside the wall perimeter; use it as a guide for placing mortar under the block's center partition and ends.

7 Butter the end of the next block with a mason trowel before you set it.

8 Holding the second block at an angle, lower it so the end and bottom seat at the same time.

11 Hit the block sharply once or twice to break the block cleanly along the groove.

12 Set the end block with the broken end facing the wall interior.

5 Lower the block over the rebar uprights and gently set it in place, aligning its face with the line on the footing.

6 Tap the block top gently to set it in the mortar. Use a level, positioned across the block, to check for level while setting the block.

9 Align the tops with a level; gently tap, rather than push, to make adjustments.

10 Break a block to fit the remaining space by tapping along a cut line to create a groove.

MIX HALF A BAG WITHOUT MESS

If you only need half a bag of cement, there is an easy way to accurately cut bags in half without any mess. Center the bag you want to cut over a rope so the rope lies under the bag. Break the top of the bag along the halfway line with a shovel or trowel. Lift the bag by pulling on the ends of the rope, and finish cutting the bag in two.

COMPLETING THE WALL

Begin the next course with a half-block, which will offset the vertical joints and make the wall much stronger. The process for setting blocks on the second course is the same as on the first. Spread mortar on top of the first course ❶. Then set the half-block in place, rough-end in ❷. Scrape away the excess mortar from the joint ❸ and use a trowel to tap the block into position ❹. To keep the wall and work area tidy, most block and brick masons will repeatedly scrape excess mortar from the joints as they set and adjust a block.

As you work, frequently check the wall for straightness. Hold a long level against the face. If a block is out of place, simply tap it into position using the level as a guide ❺. Also check the wall for plumb at the ends and along the face ❻.

After the top course is set, fill the cells with concrete and smooth the top ❼. For best results, let the wall set overnight before installing the capstones. If you haven't already done so, remove the form boards from around the footing ❽.

1 Butter the top edges of the first row of blocks with a 1-in.- to 2-in.-thick layer of mortar.

WHAT CAN GO WRONG

A concrete mixer is not the same as a mortar mixer. If you try to mix mortar in a cement mixer, two things will likely happen. First, mortar dust and sand will stick to the sides of the barrel when you add water and the mortar will not mix thoroughly. Second, gravel from the concrete will contaminate the mortar, and it will be difficult to work with. For the best results, hand-mix the mortar in a clean wheelbarrow. Add the dry ingredients first; then, in small increments, add water until the mix is a stiff milkshake-like consistency.

4 Tap the block gently with the butt end of a mason trowel to level it.

7 Fill the block cells with rubble and leftover mortar or cement.

2 Use a half-block to begin the second row. This will stagger the first- and second-course joints.

3 Scrape away excess mortar, troweling upward and toward you to remove the mortar without smearing.

5 Continue laying the second course. A long level is handy for aligning blocks along the wall face.

6 Check the wall for plumb by using a level on both the ends of the wall and at several points along the wall face.

8 Remove the form from around the footing and allow the wall to set for 24 hours before continuing work.

SHAPING, SETTING, AND GROUTING CAPSTONES

While waiting for the block to set, shape the capstones. Mark the outline of the wall on the ground and use the lines as a guide to shape the stones. Include a variety of stones: some should span the width of the wall; others can break in the middle ❶.

➔ **For more on shaping and setting capstones, see pp. 234-235.**

Once the wall is stable, dry-fit the capstones on the top of the wall ❷. Continue to shape the stones until the joint gaps are even and there is a consistent overhang. When you are satisfied, take the stones off the wall and arrange them on the ground in the same sequence.

To set the caps, mix a batch of wet mortar consisting of one part portland cement, three parts masonry sand, and water. Wet mortar makes setting the stones slightly more difficult, but because it is less porous than dry mortar, it will help keep the moisture out of the wall. This reduces the chance the wall will be damaged by freezing water during cold weather. It also helps prevent efflorescence (unsightly leaching out of salts).

➔ **For more on efflorescence, see p. 73.**

Spread mortar about 1 in. thick for the capstones. Slope the mortar back from the edge to minimize squeeze-out ❸. Check each capstone to ensure that it is level and has the correct overhang ❹. When setting stones side by side, pack mortar to the first stone and create a groove ❺. This provides a place for the mortar to go when the second stone is set. If you find there is a lot of mortar pushing up in the joints between the stones, make the groove bigger. As you set stones, check the overhang frequently and level the stones across the wall ❻. Once you've set a few feet of capstones, use a long level to align the tops and edges ❼. No matter how careful you are, there will be a little mortar squeeze-out under the stones. With a trowel, cut this mortar flush to the wall and let it fall away ❽.

After the stones are set, rake the joints at least half free of mortar to make room for the grout. For these capstones, we used a dry grout and set it with a pointing trowel. By the time you finish with the stucco (see pp. 172-173), the grout should be ready to be scraped again with the pointer trowel for a finished look ❾.

➔ **See "Dry-Mix Grouting," p. 114.**

1 **Shape the capstones on the ground first. Vary the pattern by sometimes using a single stone to span the width.**

4 **Adjust the stone to achieve the correct overhang. Remember that the brown coat and stucco can add up to 1 in. of buildup to the wall.**

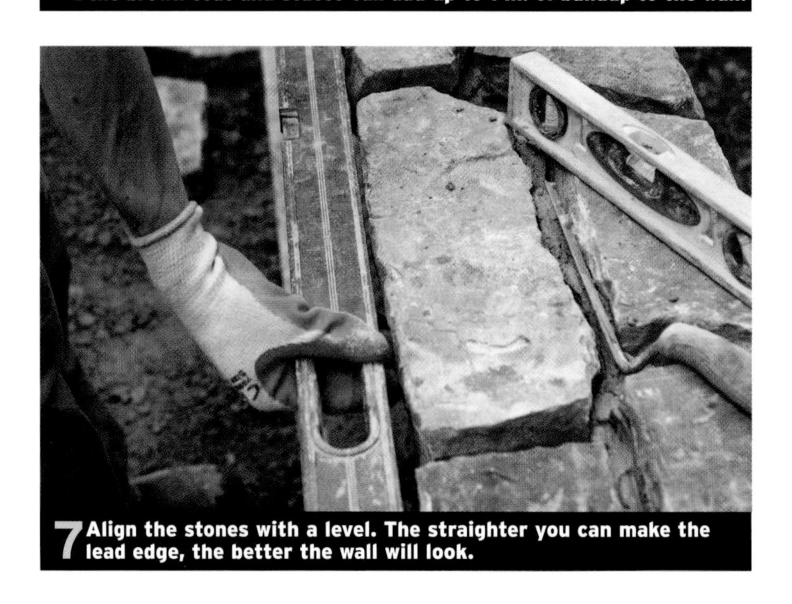

7 **Align the stones with a level. The straighter you can make the lead edge, the better the wall will look.**

2 Dry-fit the capstones on top of the wall and make any final adjustments to their shapes.

3 Spread mortar for the first stone about 1 in. thick; bevel it away from the edge to avoid squeeze-out.

5 Cut a channel in the mortar between the capstones to prevent the excess mortar from squeezing into the joint.

6 Setting capstones in wet mortar can be tricky. Continue to check the stones for level as you progress.

8 Cut way excess mortar that squeezes out from under the stone. Discard any mortar that falls to the ground.

9 Use a dry mix to grout the joints. Let the grout set until somewhat dry; then scrape the joint flush with the top of the stones.

APPLYING THE BROWN COAT AND STUCCO

Stucco is a durable, cement-based coating frequently applied to both masonry and nonmasonry surfaces. Try to apply stucco in mild weather (50° to 80° F) and when the surface is in the shade. Direct sun can dry the stucco too quickly and result in cracking. Stucco can be applied as a three-coat, two-coat, or one-coat system, depending on the type of stucco material used and the type of surface to which it is applied.

The brown coat

For this project, the block surface serves as the scratch coat, so only a brown coat and a finish coat are needed. The brown-coat dry ingredients are a mixture of one part portland cement and three parts masonry sand, bonding agent, and water. Add the bonding agent to the water before mixing the brown coat ❶ to help the finish coat adhere better. You can purchase a bonding agent when you buy the stucco. Bonding agents are usually mixed with the water for the brown coat. However, the bonding agent you use may be different, so read the mixing instructions before mixing the brown coat.

To spread the brown coat, use a brick trowel to generously load the stucco trowel (start with smaller portions until you get the hang of it) ❷. Position the edge of the stucco trowel against the bottom course ❸. Tilt the trowel to the wall while sweeping in an upward motion. If you are doing it right, the brown coat will spread evenly, without a big mess ❹. Make a few swipes, overlapping each pass. The harder you press, the thinner the layer. Typically a brown coat is 3/8 in. to 1/2 in. thick.

Cover the front and back of the wall with brown coat first, smoothing inconsistencies as you go ❺. Then apply brown coat to the ends of the wall, and finish the corner ❻. Lastly, smooth the brown coat under the cap ❼. After you have applied the brown coat, let it dry for one day before applying stucco.

Stucco

Typical stucco finish comes premixed in 5-gal. buckets and is ready to apply. Use a plastic trowel to apply it in the same manner as the brown coat ❽. Stucco mix is a little wetter than brown coat mix, so it will typically go on a little thinner. Pay special attention to the finish under the cap, because this is an area that is harder to reach with the sponge trowel ❾.

Immediately after applying the stucco, use a foam trowel to knock down, or slightly smooth, the texture ❿. Take special care at the corners to keep the finish even: The eye will be drawn quickly to irregularities at the corners ⓫.

After you finish applying the stucco, rinse all the trowels with water. Dry stucco finish is very difficult to remove. Let the wall stucco dry for at least one day before you backfill the footing trench with soil or gravel.

1 Add a bonding agent to the water before mixing the brown coat. This ensures that the stucco will bond to the brown coat.

4 To apply an even coat, tilt the trowel while making an upward stroke.

7 Take time to smooth the brown coat under the capstones. Any rough spots here will be visible through the stucco finish coat.

10 Use a foam trowel to create a texture in the stucco. Make small circular motions and be careful not to gouge the stucco.

2 Load a stucco trowel with a healthy amount of brown coat. You can start with smaller amounts until you get the hang of it.

3 Begin applying the brown coat near the bottom of the wall. If the brown coat slides off your trowel, the mix is too wet.

5 Spread brown coat to the edge. Avoid buildup at the ends, or it will be more difficult to create straight corners.

6 Apply brown coat to the end of the wall in the same manner as the front and back. Touch up the corners.

8 Apply the stucco finish coat with a plastic trowel with the same technique as for the brown coat.

9 Use a downward stroke to finish the stucco under the capstones for the best results.

11 Texture the corners by aligning the trowel's edge with the corner's edge and then pulling away from the corner.

WHAT CAN GO WRONG

If rocks or pebbles get into the mix, you'll find them when you go to spread the brown coat. Each pebble will cause the trowel to jump and leave a raised line in the brown coat. To get a smooth coat, you'll have to dig out each pebble and smooth the surface again—a process that will drive you crazy after only a few pebbles. Keep a clean mix by washing all tools, containers, and mixers before working.

BRICK WALKWAY

It is relatively easy for a novice to achieve professional-looking results when laying an unmortared (dry) brick walkway. There's nothing technical about it as long as the grade is not steep or irregular. As a bonus, laying a brick walkway is much the same as laying pavers or most unmortared stone, so you can take your pick of materials and patterns and still use this chapter as an installation guide to create your own custom walkway. Similarly, the steps of marking and preparing a solid bed are the same for walkways as for patios and paths. Best of all, the work can be done in small batches over time, as suits your schedule.

Reused brick is an excellent material choice for unmortared flatwork. It has the patina of age, lays quickly, and in many areas of the country can be found for free or at nominal cost if you're willing to haul it away.

PRACTICAL CONSIDERATIONS

Before you start your project, make sure you have enough brick on hand. Brick sizes vary, so there is no set formula for calculating how much area the brick will cover. To estimate quantity, lay down a 2-ft. by 2-ft. square using your bricks, and count them. Divide by 4 to calculate an approximate bricks-per-sq.-ft. ratio. In the event you don't have enough bricks, take a sample to your local masonry supplier and ask for help matching the color and size. Keep in mind that an exact match is not necessary. As long as you can find something close, you can weave it in with the bricks you have.

Using recycled brick

Used brick can be a great way to reduce the cost of a walkway or patio, but plan on spending some time chipping off old mortar with a hammer. Even if you want a vintage look, remove enough mortar to square up the bricks. The classifieds, craigslist℠, and even your local masonry contractor or building recycling center are all good places to check for used brick and other discounted paving materials.

Adding steps and borders

It doesn't take much of a slope to make adding walkway steps a good idea. Excessive slope makes for treacherous walking during wet or freezing weather. Even when it's nice out, it's no fun to feel as though you're trudging uphill. In addition, a gentle slope decreases the force of rainwater as it flows down the walk, which will reduce the risk of erosion or wash-out during heavy rains. In general, ¼ in. per foot pitch (slope) is recommended.

If a walkway borders landscaping elements, such as a lawn or flowerbed, consider setting the perimeter bricks in mortar. A mortared border will prevent the walkway bricks from spreading (shifting) and the edge bricks from sinking. Adding a mortared border won't change the look of the path, and it doesn't cost much more.

Excavating

Deciding whether to hire an excavator or to excavate by hand may not only depend on how many shoveling calluses you're willing to earn.

Recycled from an old chimney, these bricks have plenty of character. Character, however, was not entirely free. It took a couple of crew members several hours to clean enough bricks for the walkway.

A mortared border is easy to install and helps to keep the field bricks from spreading over time.

Deciding where to put steps before excavation will save you time in the long run.

If you have shrubs and garden beds that can't be disturbed, rule out any heavy equipment, such as a backhoe. For medium-sized jobs, skid-steer loaders can do most of the heavy lifting and can often work in the footprint of the walkway or patio. For small jobs, excavating by hand is often the only practical option.

See "Excavation and Earth-Moving Tools," pp. 40-42.

Determine the elevation of your walkway before you begin. Keep the top of the walkway flush with your yard, or slightly above (1 in.) if possible. If there is a noticeable drop in elevation near the street, mark in a rough position for a step before excavating. Remember to keep a slight slope to prevent standing water. If there is no slope to the grade, make one side of the path slightly lower than the other side or build a slight crown (convex curve) into the path.

BEFORE YOU BEGIN

For any excavation, call 811 before you dig to get assistance with finding out where utility lines, such as gas, water, and electrical lines, are buried. Also, check for subsurface drainpipes, cisterns, and tanks. Don't assume that because you're digging only a few inches down that you're safe. This is especially true if you plan to use excavating equipment.

After getting the all-clear, roughly lay out the walk to calculate the amount of materials you'll need, whether or not to incorporate a step, and where to locate it. If there will be a step, plan on putting it where the elevation changes most drastically.

Protect existing grass by working on top of plastic or plywood. Protect shrubs and flowers by covering them with plastic. If small plants and shrubs are in the way, temporarily move them. (You want to avoid doing what one homeowner I know did: remove an existing concrete path with a backhoe, destroying many plantings in the process.)

See "Protecting the Surroundings," p. 75.

WHAT YOU'LL NEED

- 15% more bricks than the walkway area
- 5 bags portland cement
- Cement blocks (for step)
- Concrete mix (for step)
- Landscape fabric
- Soil pins
- Crushed gravel (1 ton per 100 sq. ft. of patio)
- Quarry dust or river sand (1 ton per 100 sq. ft.)
- Masonry sand
- Gas-powered plate compactor
- Hand tamper
- Wheelbarrow
- Square shovel
- Masonry tools

EXCAVATION

Establishing a rough layout for excavating a straight path is easier than for a curved path (though a straight path is not always as attractive). Drive stakes at each side of the walkway, tie mason twine to them, and adjust the stringlines so they are 90 degrees to the house or porch. You can use a framing square ❶ to find the angle, or establish square using the 3-4-5 rule. Extend the stringlines the entire length of the walk. Make sure they are parallel and secure them to stakes at the street end of the path ❷. Once the stringlines are set, mark reference lines on the ground for excavation ❸.

For more on the 3-4-5 rule, see "Layout and Excavation," p. 160.

If the grade is irregular, level the stringline using a line level. Measure down from the twine to gauge the excavation depth in any given spot ❹.

Dig down approximately 8 in., cutting the sides straight and plumb as you dig. This will give you enough depth for 3 in. of compacted gravel, 2 in. of quarry dust (stone dust) or sand, and the brick. Even for a narrow walk, 8 in. of excavated soil is a considerable amount, so find a spot in the yard where you can use some fill or plan on hauling it away ❺.

TRADE SECRET

If a finished landscape requires you to precisely lay out the finished walk before excavating, consider using batter boards that can remain in place for the duration of the project. For more on batter boards, see p. 34.

1 Lay out for excavation by running a stringline to the street. Use a framing square to set the line at 90° to the house.

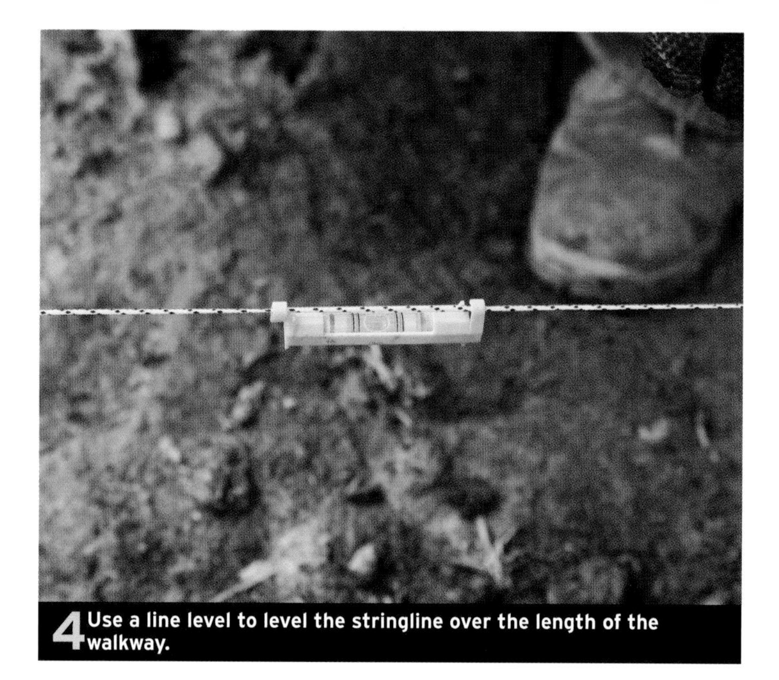

4 Use a line level to level the stringline over the length of the walkway.

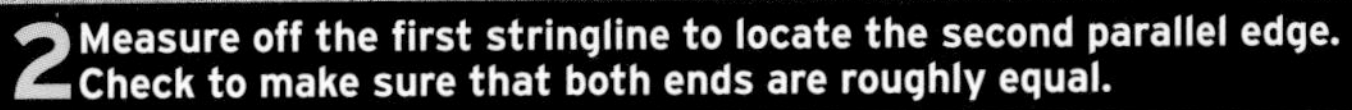

2 Measure off the first stringline to locate the second parallel edge. Check to make sure that both ends are roughly equal.

3 Mark under the stringline with marking spray. If you need to erase the line, simply scratch it out with your foot and re-spray.

5 Dig! Beg, borrow, and bargain labor from your friends and neighbors to make short work of an otherwise all-day job.

Get help with the digging

A digging crew can make short work of an otherwise all-day job. For the project shown here, a five-person crew dug the walkway bed and prepared it with gravel. They took roughly 1½ hours to do this. That's just about equal to an 8-hour day of nonstop labor for one person (or more, figuring in fatigue). Considering the work involved, it's well worth plying some friends with some pizza in exchange for a little grunt-work help.

INSTALLING THE BASE

Once you're done with the excavation, move a compactor over each area several times until the soil is hard and smooth. The longer you pack, the less chance settling will be an issue later ❶. Then spread landscape fabric over the soil to discourage weeds from growing through the walkway ❷. Fold the fabric up the sides of the excavation, secure with soil pins, and cut it just below finished grade. Spread gravel over the landscape fabric and rake it to a depth of 3 in. ❸. Then repeatedly run the compactor over the gravel until the bed is hard and smooth ❹.

➔ **See "Soil Compactors," p. 45.**

If your walkway has a step, install a footing and concrete block for it now.

➔ **See "Installing a Step," p. 186.**

Next, add quarry dust (or "fines") to a depth of 1 in., rake the surface smooth, and compact the entire area again ❺.

Now you're ready to set up new stringlines to mark the elevation on the borders of the walkway. Unlike the reference lines used for excavation, the layout stringlines need to be as accurate as possible, so take the time to position them correctly. If you have a long walkway, or a complicated layout with multiple steps and turns, it's worth the cost of renting a transit or laser level to really dial in the layout elevations. With a brick, check the elevation of the compacted quarry dust against the stringline. There should be a 1-in. gap between the top of the brick and the twine ❻ to allow room for mortar under the perimeter bricks and sand under the bricks in the field.

1 Run a soil compactor over the bed area repeatedly and in different directions. Take special care to compact the perimeter thoroughly.

4 Run the compactor over the entire graveled area several times. Take several extra passes around the perimeter.

TRADE SECRET

If you don't have soil pins to keep the fabric from falling down the side, use the end of a shovel to press the fabric into the dirt. When you remove the shovel end, the fabric should stay put. This method is not as secure as using soil pins but will hold the fabric in place at least until you can add the gravel.

2 Fold the landscape fabric up the sides of the excavation and cut it off just below the finished grade.

3 Spread a 3-in. layer of gravel over the entire path and smooth it with a garden or landscape rake.

5 Spread 1 in. of quarry dust over the entire surface and compact it in the same manner as the gravel.

6 Check the elevation of the compacted quarry dust against the layout stringline. There should be a 1-in. gap above the brick.

TRADE SECRET

Use a level or a straight board to screed the fines flat under the string. Irregularities along the edges can throw off the overall flatness of the walkway, so it's worth making the effort to get it right.

ADDING A MORTARED EDGE

A mortared edge stabilizes a walkway just as a foundation stabilizes a building. It keeps the bricks or pavers from spreading and contains the quarry dust or setting sand, reducing erosion. You can pour a narrow footing (with rebar) to give the edge a solid foundation, but in this case we decided not to because the walkway was sufficiently compacted. Setting the edge in mortar would be adequate. To set the mortared edge, spread a 2-in.-deep layer of mortar (see "Mortar Recipes," p. 61) along the edge of the walkway 4 in. to 5 in. wide ❶. The mortar should be thick enough to hold its shape when you make a channel down the middle. Position the corner brick and gently tap it into the mortar with a rubber mallet ❷. Align the brick's outside edge as close as possible to the twine without touching it.

Continue setting bricks along the same row, spreading mortar for a few bricks at a time. Each brick should set about ½ in. into the mortar without displacing the previous brick just set ❸. If a brick sinks into the mortar on its own, or if a set brick moves while setting the next brick, the mortar mix is too wet. As you work, periodically check the bricks for level ❹. If the walkway has a slope from side to side, use a mason level to align the brick tops.

When the two sides of the mortared edge are set, use a mason level to screed the quarry dust ❺. Add quarry dust to fill any depressions; use a hand tamper to compact it rather than a motorized compactor. A motorized compactor may shake the bricks you've already set out of position ❻.

With the surface between the edge bricks flat and compacted, tie mason twine across the end of the walkway at brick height ❼. Spread mortar and set the walkway's end border in the same manner as you set the side borders ❽.

The final step before setting the field bricks is to smooth the mortar base. With a trowel, bevel the mortar along the inside edges of the perimeter bricks and remove excess mortar ❾. The bevel provides room for quarry dust and field bricks to be positioned right up to the edge bricks without mortar getting in the way.

1 Spread a 4-in.- to 5-in.-wide bed of mortar under the layout stringline. Groove the top to make it easier to set the brick.

4 Check for level across the bricks as you set them.

7 Tie mason twine between the end bricks, aligned with the brick's top edge. Spread mortar and set the end bricks.

WHAT CAN GO WRONG

While the bricks should be aligned with the string, sometimes the string will get hung up or pushed by a misaligned brick. To double-check, place a longer level down the row of bricks to make sure the string is still an accurate guide.

TRADE SECRET

Stakes aren't always the best way to secure a string. If the ground won't allow you to drive a stake or you don't have one at hand, simply wrap the string around a few bricks to hold it fast.

2 Set a brick in the mortar by tapping it with a rubber mallet until the top edge is aligned with the stringline.

3 Continue setting the edge bricks in mortar. The mortar thickness should allow each brick to set about ½ in. into the mortar.

5 Screed the quarry dust between the edge bricks. Add more quarry dust if necessary to create a slight crown.

6 Compact the quarry dust before spreading mortar, using a hand tamper rather than a motorized compactor.

8 Use a mason level to align the tops. Lay it across several bricks at once, pressing or tapping gently until the tops are flush.

9 Keep the mortar bed out of the way by beveling it away from the bottom inside edge of the bricks with a hand trowel.

A BORDER WITHOUT MORTAR

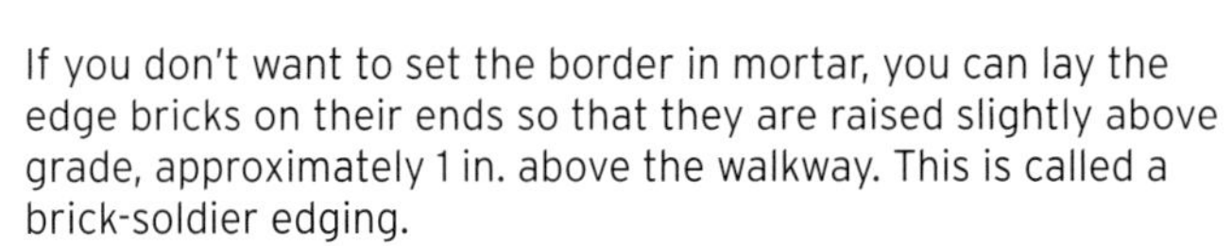

If you don't want to set the border in mortar, you can lay the edge bricks on their ends so that they are raised slightly above grade, approximately 1 in. above the walkway. This is called a brick-soldier edging.

Laying the field bricks on their ends, butted against undisturbed soil, will bind them together. If the edge bricks aren't secure, they are prone to sink. If one brick sinks, it creates a domino effect and adjacent bricks will sink as well. The edge is like a foundation. As long as it is secure, the field will be, too.

SETTING FIELD BRICKS

For this project, we chose a running-bond pattern, meaning the joints in neighboring rows are offset by the length of half a brick. All the field bricks are set parallel to the end border. Once you develop the rhythm of laying bricks, this process will go fairly quickly.

➔ **See "Brick," p. 16.**

To begin, spread an even layer of loose quarry dust under the bricks to be set. Run full bricks for the first course, and then start the first brick for the second course half a brick away from the edge bricks ❶. When you first rest a brick on the quarry dust, the top should be about ½ in. high. Strike the brick three or four times with a rubber mallet to set it flush to the tops of adjacent bricks ❷. If the brick won't sit flush with the other bricks, scoop out some of the stone dust beneath it. Then reset the brick. It's easier to gauge the correct amount of quarry dust if you prepare an area for several bricks at once. For best results, hold a long mason level or straightedge over several bricks and the mortared borders as you set them ❸. For a wide walk or for a walk that does not have a natural slope, set the bricks with a slight (½-in.- to ¾-in.-high) center crown to prevent standing water from accumulating later.

Cut half-bricks at the ends of every other row as you need them, or a few at a time ❹. Leave a ⅛-in. gap between the bricks so the joints can be filled with sand after all the bricks are set.

➔ **See "Cutting Tools," pp. 38-39.**

The best brick setters develop a rhythm that is not too fast and that wastes very little energy. In general, work toward yourself rather than kneeling on the brick and working away from yourself. Keep the area neat and free of extraneous tools and debris ❺. A helper can be invaluable, providing a continuous supply of bricks (including cut bricks), quarry dust, and even drinking water.

After every three or four rows, check for straightness with a mason level or straight board ❻. Periodically measure from each side of the row to the end to be sure that one side is not creeping ahead of the other. It's especially important to check if two people are setting bricks side by side.

Filling the cracks

Once the bricks are laid, fill the joints halfway with sand. Make sure the sand you use has been screened multiple times so that no pebbles remain and all the small particles will fall into the joints. Masonry sand should work just fine. Then run water over the walkway to remove air pockets and to pack the sand down. Continue sweeping sand into the joints until they are full. Then run water over the patio again, letting the sand settle ❼. Leave a little sand nearby because you might need to add more after a few rains.

1 Start the second row half a brick away from the border to form a running-bond pattern. Leave a ⅛-in. gap between bricks.

4 Cut the end bricks for every other row as you progress. If using a power saw, brace the back of the brick to secure it during the cut.

TRADE SECRET

Another way to flush the brick tops as you set them is to use a short board. Place the board so it spans the brick to be set and the adjacent brick already set to the proper height. Hit the top of the board until the bricks are flush. This method is also useful if you don't have a rubber mallet and don't want to mar the brick faces.

2 Strike the brick with a rubber mallet to set it flush to the adjacent bricks.

3 Use a mason level to align the brick tops. While gently pressing down on the level, tap the brick until the tops are flush.

5 Work toward yourself, keeping unnecessary materials and tools out of the way.

6 Check every other row for straightness. Hold the level against the leading edge and make small adjustments as necessary.

7 Pour sand over the entire surface and work it into the joints with a broom.

TRADE SECRET

A good way to get sand down into the cracks is to use an impact driver or roto-hammer to vibrate a board set on top of the bricks. This quickly settles the sand into the cracks. Don't use a motorized compactor or you run the risk of breaking bricks.

INSTALLING A STEP

To prevent a step from moving over time, it's a good idea to give it a solid concrete foundation. The foundation described here is a simple earth-formed concrete slab on three sides with a fourth wall made from a 2-in. by 8-in. board held in place by a couple of concrete blocks. We added a rebar ladder to give the footing strength ❶. Fill the form with concrete (for a job this small, you can mix it from a bag) and allow it to set overnight before laying the concrete block for the steps.

See also "Pouring a Concrete Footing," p. 162, and "Installing a Footing," pp. 191-193.

On top of the footing, install two rows of concrete blocks (different sizes for each step) using the same mortar recipe as for the mortared border ❷. Outdoor steps tend to have much deeper treads than stairs inside a house, so a comfortable width spaces the block faces 13 in. apart for a finished tread width of 14 in. (which includes 1 in. for tread overhang) ❸.

For more on calculating step weight and depth, see "Designing Stairs," p. 137.

For more on creating a block structure for steps, see "Making a Mock-Up," p. 142.

The riser is formed from bricks set on end (also call a soldier course) set directly on the block. To make quick work of laying a soldier course, set the first and last brick and run a stringline between them, aligned with their top outside edges. Fill in with bricks to the twine to achieve uniformity ❹. For the step treads, we matched the existing 2-in. fieldstone treads that lead to the porch ❺.

See also "Brick-Veneered Steps," pp. 131-133.

See also "Setting Stair Treads," p. 149.

1 Form a simple step footing by digging into the bank and bracing a board against the down-slope side.

4 Set bricks on end to form a soldier course. Set the end bricks first, then run mason twine between them to serve as a guide.

2 Set concrete block on the footing to achieve the proper height for setting the stair treads.

3 Position the blocks so the lead edges are 1 in. less than the tread depth.

5 Set the treads in dry mortar, then remove the excess mortar from between the joints. Allow to set for 24 hours before grouting.

DO YOU NEED A STEP?

To determine whether you need to add a step to your walk, you need to know the overall length and drop. The maximum drop for a comfortable walkway is ½ in. per foot, so multiply the overall distance in feet by ½ in. For example, if your walk is 10 ft. and the overall drop is 5 in., you don't need a step. The walk shown here was roughly 20 ft. long with a 16-in. drop. We could have added a 6-in. step with 10 in. of slope. However, that's not a very relaxing approach to a front porch. Instead, we added two 5½-in. steps and reduced the slope to ¼ in. per ft. This resulted in a comfortable approach to the porch with an opportunity to pause and enjoy the landscaping.

MAILBOX COLUMN

BUILDING A STONE COLUMN WITH a built-in mailbox is a particularly rewarding project because you'll see it every time you pull in or out of your driveway. The design shown here features an optional newspaper holder. The project can be completed in just a few days, and all of the techniques described here can help you tackle bigger and more complex mortared-stone projects.

If you prefer, the mailbox and newspaper slot can be omitted if you just want to build a purely decorative column. The steps covered in this chapter include pouring an earth-formed foundation, setting concrete block, applying stone veneer, and setting mortared capstones.

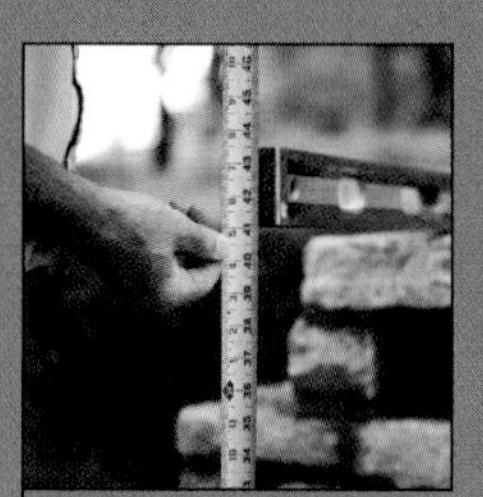

GENERAL PREPARATION

INSTALLING A FOOTING

SETTING THE BLOCK CORE

ADDING STONE VENEER

SETTING CAPSTONES AND GROUTING

451

PRACTICAL CONSIDERATIONS

When deciding where to locate the column, first contact your local postal service, Department of Transportation (DOT), and property owner's association to determine any right-of-way restrictions. The restrictions for the column shown here included a 9-in. setback from the edge of the road and a 41-in. to 48-in. mailbox door height, as measured from the pavement.

Whatever mailbox you choose will literally be set in stone. If you're going to use an old mailbox, make sure that it is rust-free and in good working order and that it suits the style of stonework you've chosen. For our column, we selected a standard black rural box that wouldn't detract from the formal stonework and wrapped it in self-adhesive flashing to prevent rust.

Hit the mark. Find out ahead of time where and how high you need to build your column and then build to those specifications.

Stay visible, stay safe. An orange cone is an easily recognized signal to alert drivers to pay special attention.

When working close to or in the road during a project, you may need to request a temporary road closure permit from the DOT. You may also have to request a temporary sidewalk closure permit from your local municipality. Check on these requirements well in advance. If you are working on a quiet street, with an off-street area to stage materials, a few orange traffic cones to alert oncoming drivers should suffice.

BEFORE YOU BEGIN

If the stoneyard is delivering materials, schedule the delivery for the day before you plan to do the work. If you have a large pickup truck, this job is small enough that you can pick up the stone yourself. Plan on making several trips if you have a small pickup. Even the small amount of material required for this job can easily overwhelm an undersized vehicle.

With a helper, you can expect to complete the mailbox in two or three days. To ensure full days of work, however, it's a good idea to gather all the materials and store them at the site beforehand. In addition, locate a water source or arrange for a way to get water. You're going to need about 100 gal. for this job.

➔ **See "Establishing Staging Areas," p. 77.**

Always stage materials as close to the job as you can get. For this job, the masonry-supply yard delivered the stone and the crew brought everything else.

WHAT YOU'LL NEED

- 1 ton granite building stone
- 1/2 ton masonry sand
- 4 12-in. x 12-in. concrete blocks (more if your column is taller)
- 8 bags portland cement
- 6 to 8 bags concrete mix
- 6 to 7 fieldstones for caps
- 1/8 ton clean gravel
- 4 10-ft. lengths 1/2-in. or 5/8-in. rebar
- Plumb bob
- Nylon mason twine
- Plywood
- Wheelbarrow
- Square shovel
- Masonry tools
- Mason level
- Tape measure
- 1 can marking spray

LAYOUT AND EXCAVATION

A 2-ft. by 2-ft. stone column has a 12-in. by 12-in. concrete-block core and a 3-ft. by 3-ft. footing, or foundation. When marking the dimensions, be precise. Accurate dimensions take the guesswork out of digging and ensure that the column is located correctly with the proper setback. If you're using the old mailbox to position the new column, mark the location before removing the old column ❶.

An earth-formed footing is a fancy way to say that you're using a hole to contain the concrete instead of using form boards. It is a perfect method for a small project like this where drainage is of limited concern and there is no structure to tie into ❷. In locations that experience extreme cold, the column may require a deeper footing. Check with your local building department for the requirements in your area. If you're building the column near lawns and garden beds, lay down plastic nearby where you can toss the excavated soil. You will need some of the soil to backfill around the column; the rest will have to be moved elsewhere. When you've reached the proper depth ❸, square up the sides and bottom, remove loose soil, and compact the remaining soil with a hand tamper to prevent unsightly settling.

➔ **See "Soil Compactors," p. 45.**

A 12-in.-deep hole will allow for 2 in. of clean gravel at the bottom for drainage, 8 in. of concrete, and 2 in. of topsoil to cover the footing when the column is finished. Compact the gravel in the same manner as the soil. If you don't have a soil compactor, use the flat side of a concrete block to tamp the soil and gravel ❹. Let the weight of the block do the work. If you push or slam the block too vigorously, it might break.

1 Locate the column footing accurately. Take the time to precisely mark the setback from and alignment to the road.

2 Excavate the soil. Soil adjacent to the road is typically very compact and you may need a mattock to loosen it.

3 Dig the footing at least 12 in. below grade. Place a level across the hole to check the depth as you progress.

4 Compact 2 in. of gravel at the bottom. For this small area, a concrete block can serve as a tamper.

ADDING REBAR AND CONCRETE

Adding rebar to the footing with vertical ties to the column may seem like overkill, but with the column so close to the road, a car or passing snow plow could easily clip it. Without rebar, the chance is much greater that it could be knocked from its footing. For the rebar, you will need eight pieces of ½-in.- or ⅝-in.-dia. rebar, cut 32 in. long to create the footing grid. You'll also need two pieces 3 ft. long, with a 1-ft. 90-degree bend on one end to tie the footing to the column. You can rent a rebar cutter/bender from your local rental store ❶.

Cut the rebar at least 2 in. shorter than the width of the footing ❷ and use rebar chairs to elevate the rebar 2 in. to 3 in. above the gravel ❸. A footing this size will require six to eight 100-lb. bags of concrete mix (three batches in a wheelbarrow), so it's probably not worth the expense and hassle of renting a concrete mixer. To save the lawn, use buckets to transport the concrete from the wheelbarrow to the footing hole ❹.

➔ **For more on cutting and bending rebar, see p. 68.**

You can tie the vertical supports to the rebar grid before adding the concrete, but it's easier to work them into the concrete for a small project like this ❺. The footing surface needs to be level, but it doesn't need to be super smooth because soil will eventually cover it ❻. Let the concrete cure for at least a day before laying block and stone.

TRADE SECRET

It's easy to cut rebar with a grinder or circular saw fitted with a metal-cutting blade. It's also fairly easy to bend rebar by stepping on one end and pulling up on the other. However, it's a lot more difficult with the ⅝-in. rebar used here, so you might want to rent a rebar bender.

1 You can rent a rebar cutter for the day, but be forewarned that cutting ⅝-in. rebar with this tool requires significant leverage.

4 Once the rebar is set, pour the concrete into the footing hole.

2 Tie the rebar with wire ties to form a grid; space the rebar 6 in. to 8 in. apart.

3 Place chairs under the rebar at the corners to keep the rebar in the middle of the footing where it will provide the most strength.

5 Locate the center of the column and work the longer sections of rebar into the concrete, L-end down, until they contact the grid.

6 Smooth the top of the footing with a trowel or square shovel and check for level.

ALIGNING AND SETTING THE FIRST BLOCK

With a 24-in.-wide stone column, a 12-in.-square block leaves room for 6-in.-thick stone veneer on all sides. To begin, place the first block at the center of the footing, before spreading mortar, and check your setback and alignment to the street. If the footing placement was slightly off, you can correct it now by repositioning the block. It's not critical that the rebar protrude exactly in the center of the block. However, avoid positioning your column too close to the edge of the footing. Once the block is properly positioned, trace its outline on the footing and set it aside for a moment ❶.

To set the concrete blocks, mix a batch of mortar consisting of half a bag of Type S masonry cement and nine shovels of masonry sand. Add water in small amounts until the mortar is the consistency of a milkshake. Spread mortar 1 in. thick under the entire area where the block will rest on the footing ❷. When setting the first block, gently lower it over the rebar until it rests on the mortar ❸. Check for level, and use your trowel handle to tap the block into position, if necessary ❹. Remove excess mortar around the base of the block and spread it on top.

1 **Locate a concrete block, checking for setback and alignment, and mark its position on the footing.**

2 **Spread a 1-in.-thick bed of mortar for the first block. Well-mixed mortar should be thick enough to hold its shape.**

3 **Set the first block without pushing it down. You don't want to squeeze out the mortar underneath it.**

4 **Level the block in both directions. Adjust it by tapping rather than pushing.**

ADDING BLOCKS TO MAKE A COLUMN

Going from a single block to a column is simply a matter of adding more blocks. It is important, however, that you check the growing column for plumb frequently and make small adjustments as you progress. To start add, or "throw," enough mortar to make a 1-in. bed for the next course ❶. Set the next block in the same manner as the first and tap it into place ❷. Scrape excess mortar from the joints and spread it on top for the next block ❸. As you lay each block, check that the top is level and the sides are aligned and plumb ❹. Repeat this process until the height of the block column is a few inches short of the newspaper holder height.

The last step is to insert wall ties into the block joints before the mortar dries. For this application, put wall ties between every other block on all sides ❺. Allow the mortar joints to dry overnight and fill the block cells with concrete mix and rubble the next day. Use your trowel to smooth the top ❻.

1 Throw an even 1-in.-thick layer of mortar on the block's top.

2 Set the next block and level it in both directions by tapping it with your trowel.

3 Remove excess mortar that pushes out of the joints and spread it on top for the next block.

4 Align all the blocks vertically after setting each block. If the blocks get out of plumb, gently tap them back into position.

5 Insert ties about one-third of the way into the mortar joint. The part that protrudes will tie into the stone veneer.

6 Fill the hollow center of the column with rubble and concrete after the mortar sets; smooth the top.

SHAPING STONE

Whether you are working by yourself or with a helper or two, shape the stones before you begin setting them to save time. Use a brick hammer, along with a rock hammer and chisel, to square up all the stones. Shaping the stones will give the column a clean, formal look that is appropriate for granite.

Begin by determining which side of each stone is closest to being square and smooth. Check dimensions with a tape measure for width and depth ❶. Rarely will stones be perfectly bricklike, but you can strive for that goal. Trim the stones to the general shapes you want ❷; you may have to shape them a little more before setting them in mortar, but it will save time to do as many as you can ahead of time. Shape the face of the stone, too, to give it an evenly worked look ❸. When you begin setting the column, you should have a nice pile of dressed stones ready to go ❹.

➔ **See also "Shaping Tools," p. 20.**

1 Measure a stone and check it against the space you have to fill.

2 Use a brick hammer to shape the stone to the width, height, and thickness you want.

3 Chip the face to give it an evenly worked appearance. This helps to give the column a uniform look.

4 Stack finished stones near where you will need them.

SETTING STONE VENEER

The most important aspect of building a column is to get the corners plumb and straight. In fact, you can get away with less than perfect stonework as long as the corners are dead on. To make perfect corners, cut a piece of plywood to the column dimensions—in this case, 2 ft. by 2 ft. Center the plywood on top of the block column ❶ and secure it with a concrete nail. (You can also simply use a couple of extra blocks to hold the plywood in place.) Drive a small nail into the point of each corner of the plywood. Using a plumb bob, mark the concrete footing directly below each corner ❷. Then drive a concrete nail at each mark into the footing, leaving the nail head ¼ in. proud of the surface. Use a framing square to mark a straight line between the corner nails ❸ and tie mason twine from the nails to the corners of the plywood. If the blocks are plumb, the distance from the blocks' corners to the nails will be equal ❹. Setting your corner stones to the stringlines will make your column perfectly straight.

Before you continue, you'll need to mix a batch of mortar consisting of half a bag of portland cement and 14 shovels of masonry sand. The mortar should be wet enough to stay in a ball when you throw it in the air.

➔ **See "Hand-Mixing Mortar," p. 62.**

Start at the corners

Begin by setting the stone veneer at the corners. First, spread a 1½-in.-thick bed of mortar on the footing. Then place the corner stone in the mortar and tap it into place so the corner lines up closely with the stringline (but not touching it) ❺. Set another stone at the adjacent corner. If you've done a good job shaping, the tops of these stones will be close to level ❻. Measure the space between the corner stones and fill the gap with one or two stones ❼. Then place mortar behind the stones, and use your pointing trowel to pack mortar into both the horizontal and vertical joints ❽. Fill the joints flush to the front edge of the stones and pack mortar in all voids.

If you are working alone, set the first course on all four sides of the column before moving on to the second course. If you are working with one or more helpers, keep even with each other and tie in the corners at the same time. As you work up the column, embed wall ties into the mortar as you encounter them ❾. If two corner stones are not level, make the course level with at least one of them ❿. Continue building until the stone reaches the plywood, at which point you can remove the plywood and twine.

>> >> >>

1 Place a plywood square cut to the exact column dimensions and center it on top of the concrete block.

2 Use a plumb bob to locate the spot directly below the plywood corner. Drive a concrete nail into the footing at this spot.

SETTING STONE VENEER (CONTINUED)

3 Mark straight lines between the corner nails using a framing square. This line represents the footprint of the finished column.

4 Tie mason twine from the nails in the plywood corners to the nails in the concrete.

7 Lay mortar and set a stone (or stones) in the gap between the corners. Reshape the stone to achieve a satisfactory fit.

8 Pack mortar behind and between the stones. Use a pointing trowel to solidly pack mortar into the joints.

5 Set a corner stone in a 2-in bed of mortar and gently tap it into place with a brick hammer.

6 Set the next corner stone and level the tops. If you've done a good job shaping, the corner will align with the string.

9 Embed ties between the stones. Pack mortar below the tie, press mortar into the tie, and pack more mortar on top.

10 Level each course as you progress. If one corner stone is higher, level the course to the lower corner.

ADDING A NEWSPAPER HOLDER AND MAILBOX

To form a newspaper holder, set 4-in.-high stones atop the column (in the center) to create the sides ❶. Smooth the mortar, sloping it slightly to the outside, to form the base. Use stones large enough to bridge the sides to form the top ❷.

The mailbox goes on top of the stones that bridge the sides of the newspaper holder. When setting the mailbox in a bed of mortar, level the box from front to back ❸. If there is any tilt, it should be slightly toward the door to ensure drainage. Test the door to make sure it opens freely before laying stone around it. Continue laying corners and stones around the mailbox until you reach the height of the cap ❹. Because the mailbox itself can't support weight, span the top of the box with a stone that is supported on each side by masonry ❺.

1 Build a newspaper holder by setting 4-in.-high stones to form the sides.

2 Add a top to the newspaper holder by spanning the sides with flat stones.

3 Set the mailbox in a bed of mortar. Level the box from front to back.

4 Continue to set stones and pack joints around the mailbox.

TRADE SECRET

Before you install a new mailbox, wrap it with self-adhesive flashing to protect it from moisture and rust. If you are using an old mailbox, take the time to remove any existing rust and give it a fresh coat of outdoor paint.

5 Span the top of the mailbox with a single stone to support the weight of the capstones.

SHAPING AND SETTING CAPSTONES

Using 4-in.-thick capstones can make the column look stronger and more substantial (see the photo on p. 189). To set the cap, begin by shaping four capstones so that each one has a 90-degree corner.

With your framing square as a guide, mark as close to the edge of the stone as possible ❶. Remove the edges of the stone along the cut line with a cold chisel and mallet ❷. Make any small adjustments with the edge of a brick hammer as needed.

Once you've squared up the capstones, spread a 1-in.-thick bed of mortar on the column and set the first corner perfectly level ❸. Then follow suit with the other capstones. Fill gaps with smaller stones ❹.

1 Choose a capstone that already has one corner close to 90 degrees. Mark a cut line as close to the edge as possible to avoid waste.

2 Remove excess stone with a cold chisel and mallet. Work the edge from both the top and bottom.

3 Set the cap's corners first and align them with a level. A perfectly level top will add a look of solid orderliness to your column.

4 Fill the center last. Use a rubber mallet to set the center stones, aligning the tops to the corner stones.

GROUTING

You can grout the column as you build it or after all the stones are set. The latter ensures that the joints will be a consistent color. Just be sure to scrape the mortar joints to a depth of 1 in. as you lay each course. Otherwise, the mortar may harden before you can complete the job and there won't be enough room for grouting later. If you grout after each course, take care to mix each batch with the same amount of sand and portland cement so the color is consistent.

Use a pointing trowel to pack each joint flush to the stone surface. Then scrape the joints flush with a pointing trowel or with a small wire brush. Use a whisk broom to sweep the joints after you scratch them out. Use the same technique to grout joints between the capstones.

The flush joints, combined with the granite, give the finished column a historic look.

➔ **See "Dry-Mix Grouting," p. 114.**

Scrape the joints flush with a pointing trowel to give the column a timeless look.

MANUFACTURED STONE FIREPLACE

MANUFACTURED STONE, ALSO KNOWN as cultured stone, is a great way to dress up a new or existing fireplace. It has the look and feel of natural stone yet is lightweight and easy to work with. Manufactured stone requires minimal shaping and is available with premade corner pieces. Preparation is no more difficult than applying a brown coat as you would for stucco (see p. 172). Because it's lightweight, manufactured stone doesn't typically require structural modifications.

For the fireplace shown here, we chose a manufactured stone that has a naturally weathered look.

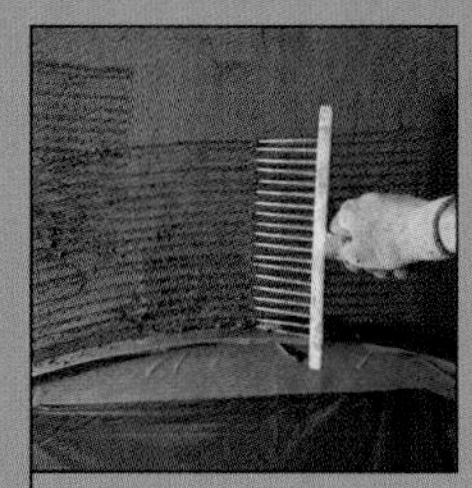

PRACTICAL CONSIDERATIONS

Whether a fireplace is faced with manufactured stone or natural stone will not affect how it functions. Natural stone, however, is heavier and the framing that supports it will typically need to be beefed up to take the load. Manufactured stone is lightweight and can be attached to the wall in the same manner as tile or marble. Most framing can accommodate its extra weight without modification. However, if you are unsure, ask a builder or an engineer.

In either case, you'll need to ensure that your fireplace unit is completely installed and in good working order before you begin. It is much more difficult and costly to add gas starters, install electrical outlets, and reposition vents or flues after the stone has been installed.

Finally, consider your mantel. Unlike the thick veneer of natural stone, manufactured stone can't support the weight of a heavy stone mantel. For this job, the builder created pockets in the wall framing to support the mantel independent of the veneer. Into these pockets, we anchored stone corbels to transfer the mantel's weight to the framing (rather than to the manufactured stone veneer). You could also use a wood mantel, which is much lighter; it is typically installed first and the stone is laid up to it.

WARNING

In some pictures you'll see the installers wearing gloves; in others, you might not. The simple explanation is that sometimes gloves get in the way. But it's hazardous to take them off, especially when working with materials like sharp-edged lath. Concerning safety, it's always best to keep your gloves on.

BEFORE YOU BEGIN

Before beginning the stonework, spread cardboard or plywood to prevent damage to floors, even if they're unfinished. Protect the trim, walls, and fireplace inserts from cement stains with masking tape. If there are nearby windows, cover them with plastic or foam board just in case a fragment gets flung at one as you're cutting the stone. In addition, cover the fireplace doors to protect them from cement stains. Lastly, check over all the surfaces to make sure that everything is ready to go, cutting any loose strings from the backerboard and trimming proud edges. Trimming backerboard later will make a mess in the mortar.

For our project, the backerboard had already been installed by the builder. If you are installing your own backerboard, it's a good idea to add a layer of roofing felt between the backerboard and whatever is behind it (framing, drywall, etc.). This will prevent the wall from absorbing moisture from the scratch coat while it cures.

WHAT YOU'LL NEED

- 25 sq. ft. manufactured stone flats
- 3 lin. ft. manufactured stone corners
- 1/4 ton masonry sand
- 4 bags Type S masonry cement
- 1 bag portland cement
- 30 sq. ft. metal lath in sheets or a roll
- Wheelbarrow
- Tin snips
- Masonry tools
- Plywood or cardboard to cover the floors
- Plastic to cover the fireplace unit
- Masking tape

Cover all nearby surfaces **to protect them from the stone, flying stone chips, and stone dust.**

Eliminate proud edges **and loose strings from the backerboard and thoroughly clean the site of dust and debris.**

CUTTING AND SECURING LATH

The first step is to install metal lath over the backerboard. You'll want the lath to cover wherever you intend to set stone. Overlap pieces by at least 1 in. To make the fewest number of cuts, measure and cut the sidepieces first ❶. Tin snips are best for cutting metal lath, but a grinder with a 4-in. metal-cutting wheel will also work in a pinch. There's no need to mark the lath before cutting. Save time by simply using your tape measure as a guide ❷. After you cut each piece, secure it to the wall ❸.

➔ **See "Wire Mesh and Lath," p. 66.**

To fit lath over an arched fireplace insert, cut a sheet of lath to cover the entire width and secure it with a couple of fasteners. Make sure that it overlaps the side lath by at least 1 in. Then cut the curve of the arch with tin snips using the front of the fireplace as a guide. This method is faster and more accurate than trying to cut the arch shape on the ground ❹. Once secured, cut the lath flush to any openings ❺ or other areas where you will not apply stone.

When securing the lath, avoid bubbles, bulges, or loose areas; they cause uneven places in the scratch coat and alter the thickness of the setting mortar. Secure the lath as close to the corner as possible. This will help keep a straight edge against which to set the stone ❻.

Before applying the scratch coat, cover the fireplace unit with plastic and masking tape ❼. >> >> >>

TRADE SECRET

Adhesive characteristics for masking tape are identified by color. Blue tape, sometimes called painter's tape, is strong enough to mask an area with paper or plastic but releases easily without peeling paint or leaving residue.

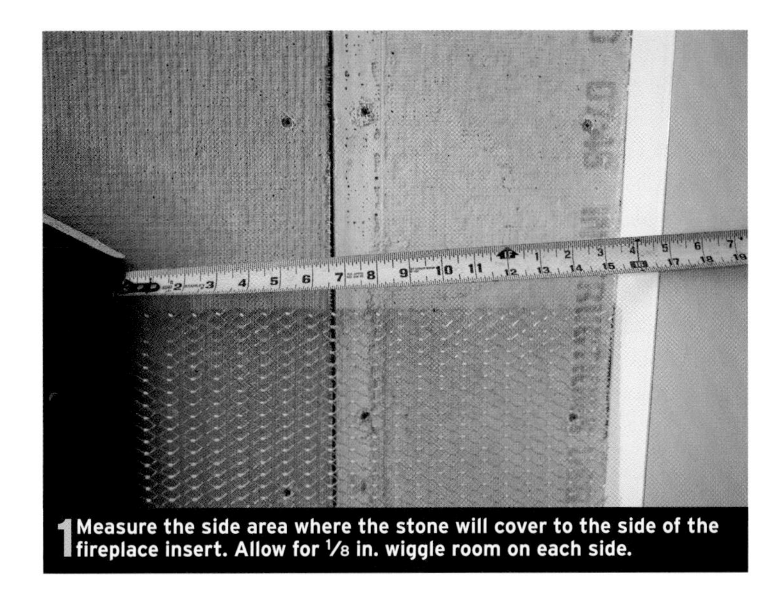

1 Measure the side area where the stone will cover to the side of the fireplace insert. Allow for 1/8 in. wiggle room on each side.

2 Cut the lath using your tape measure as a guide. Be careful; freshly cut lath is extremely sharp!

3 Secure the lath every 4 in. to 6 in. around the perimeter and every 8 in. in the field.

CUTTING AND SECURING LATH (CONTINUED)

4 **Hang lath so it overlaps the arch and make your cut before securing the bottom edge.**

5 **Trim the lath carefully around any areas that will not be covered by stone (such as the corbel pockets shown here).**

6 **Drive fasteners as close to the outside corners as possible; a solid straight edge facilitates laying stone.**

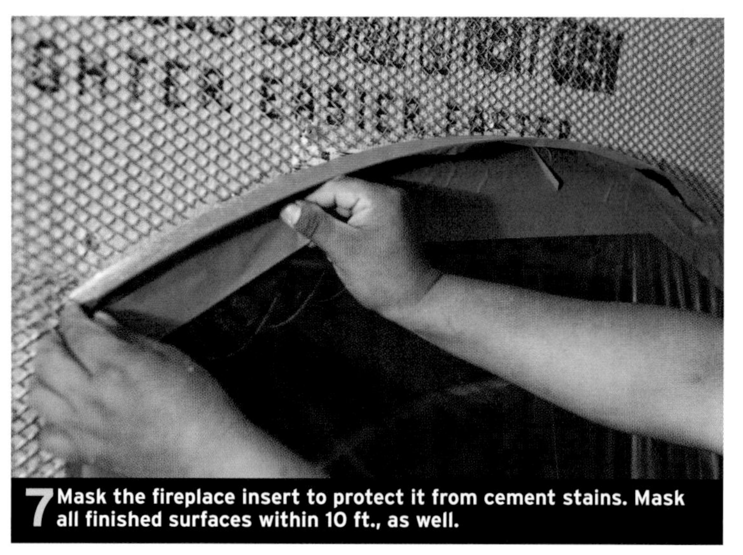

7 **Mask the fireplace insert to protect it from cement stains. Mask all finished surfaces within 10 ft., as well.**

FASTENERS FOR LATH

Use large-headed fasteners long enough to penetrate the studs behind the backerboard to a depth of at least 3/4 in. For large jobs, I use roofing nails loaded into a pneumatic nailer for efficiency and speed. For small jobs like this one, truss-head or large, flat-head screws driven with an impact driver work great. For metal studs, which are often required by code for fireplace surrounds, use screws with a self-tapping tip. You can also attach lath with staples to either wood sheathing or drywall.

For small jobs, secure lath with screws. For larger jobs, consider using a pneumatic nailer loaded with roofing nails to speed the process.

APPLYING A SCRATCH COAT

Once the lath is installed, mix 1 bag portland cement and 12 shovels masonry sand for the scratch coat. Use a brick trowel to load a healthy amount of mortar from the bucket to the stucco trowel ❶. If the mortar won't stay put on the trowel, it's probably too wet.

➔ **See "Hand-Mixing Mortar," p. 62.**

Spread the mortar evenly over the lath from the bottom up. To do this, place the bottom edge of the loaded trowel against the lath, tilt it back, and move it up at the same time ❷. If you've never done this before, it may take a little practice to completely fill the voids and cover the lath in one pass. The goal is to leave a smooth surface with no lath visible. That said, it's okay if some of the lath's pattern telegraphs through the mortar when you're done.

When you get to the edges, place the mortar closer to the edge of the trowel. You may find that a smaller trowel works better here. For both the inside and outside corners, take special care to fill any voids and completely cover the lath ❸. Apply mortar to the top of the hearth, as well ❹.

Let the mortar dry for about 30 minutes. It should be stiff but still soft enough to score. You can use the tip of your trowel, a store-bought scoring tool made for this purpose, or a homemade tool made by driving several nails, 1 in. apart, through a scrap of wood ❺. Regardless of the tool, score the mortar in one direction only. Shallow scratches are all that's required to create a good bond between the stone and scratch coat ❻.

Let the scratch coat dry for 24 hours before applying the manufactured stone. To make best use of your time, it's a good idea to get ready for the following workday by doing a general cleanup and organizing the stone for the next day. Pile the stones, according to size, far enough away from the fireplace to allow plenty of room to work yet close enough to be within easy reach ❼. >> >> >>

WHAT CAN GO WRONG

Areas of improperly secured lath will spring back after a trowel pass, creating a gap between the lath and the mortar. To fix this problem, drive an extra fastener to secure the lath and go over the area again with the stucco trowel.

1 Load a stucco trowel with a generous amount of mortar. Thicken the mortar if it doesn't stay on the trowel.

2 Spread mortar with an upward stroke. Tilt the trowel as you move it to press the mortar into the lath.

3 Load the lath at the corners first. Then smooth the mortar to create an even edge.

APPLYING A SCRATCH COAT (CONTINUED)

4 Spread mortar on the hearth surface. An even scratch coat will make laying level stone easier.

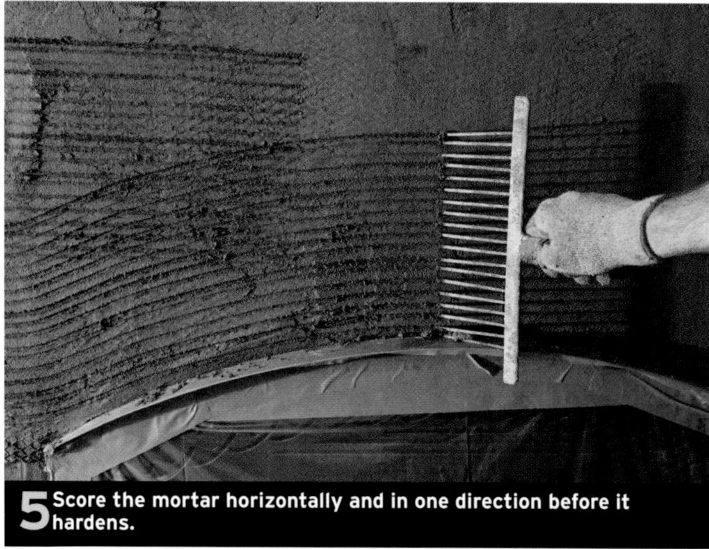

5 Score the mortar horizontally and in one direction before it hardens.

6 Avoid scoring too deeply and exposing the lath. A shallow scratch is all that's required to create a good bond.

7 Organize the stone by size and position within easy reach of the fireplace.

LAYING THE STONE

Once the scratch coat is dry but before setting any manufactured stone, you'll need to mask the floor. You've already laid protective cardboard over the floor, but if you leave it against the edge of the fireplace, it will become trapped under the first row of stone. Instead, mask the floor with strips of roofing felt, thin cardboard, or heavy-duty plastic. Tape it along the outboard edge, leaving 2 in. of exposed floor. (Note: Roofing felt can leave black residue on finished floors. Here, the floors were unfinished, so it was not a problem.) Mask the floor directly under the area where you'll be laying stone. ❶.

For manufactured stone, mix ½ bag of Type S masonry cement and 4 shovels of masonry sand. The mortar should be wet but thick enough to hold its shape when put on the back of the stone with a brick trowel ❷. Mix the mortar in a wheelbarrow but transport it into the house in 5-gal. buckets.

See "Mixing Mortar in a Wheelbarrow," pp. 62-63.

Set the corners first

Manufactured stone has an obvious textured front and a flat back. Set the corners first, from bottom to top. To begin, cover, or "butter," the back of a corner piece thoroughly and push the stone against the wall until the mortar bulges ❸. You may have to hold the stone in place for five to ten seconds before it sticks. Leave a ¼-in. to ½-in. layer of mortar between the stone and the scratch coat. Avoid letting the stone touch the scratch coat or the bond will be significantly weakened. Scrape away excess mortar and use it to butter the next stone ❹.

With manufactured stone, you can set all the corner pieces first and then set filler stones between. Corner stones have a long and a short side; alternate long and short to avoid aligning the vertical joints of two or more courses ❺. Once you have laid the corners to the top of the hearth on both sides, scrape out the excess mortar from the joints. Set the rest of the stones beginning at the bottom, adjacent to the corner stones ❻.

Butter the back of each stone and scrape away the excess on the edges so you don't smear mortar on the stones that have already been set ❼. Use both hands to set the stone in place, pushing it gently against the wall ❽. If the stone is close to position but still needs a little adjustment, tap, rather than push, it into place ❾. You will have much more fine-tuning control when you tap a stone. Continue setting stones from the corners toward the center. At the corners, match the corner stone's width when selecting an adjacent stone ❿. For common techniques on how to improve stone veneer's overall appearance, see "A Matter of Aesthetics," p. 211.

Eventually you'll need to fit stones to complete a row. Don't spend too much time looking for the perfect stone to fill a gap. It's much quicker to use the sharp edge of your brick hammer to break a stone ⓫ or use a diamond blade on an angle grinder to cut one to length.

See "Cutting Tools," p. 38.

>> >> >>

WHAT CAN GO WRONG

If the stone doesn't stick, the mortar probably isn't wet enough. Add a small amount of water and mix it in thoroughly. If the stone slides or rubs against the scratch coat, the mortar is too wet. Add equal parts sand and cement to thicken it.

1 Protect the floor with felt or cardboard, beginning 2 in. back from the scratch coat.

2 Mix mortar so it holds its shape when applied to the back of a stone.

3 Set corner stones first. Hold each stone for 5 seconds or until it stays in place.

LAYING THE STONE (CONTINUED)

4 Remove excess mortar before placing the next stone.

5 Alternate long and short sides of corner stones to stagger joints

6 Fill in stones from the corner to the center (and from bottom to top).

7 Butter stones generously and remove excess mortar after setting each stone.

8 Tilt each stone into place and align it with adjacent stones.

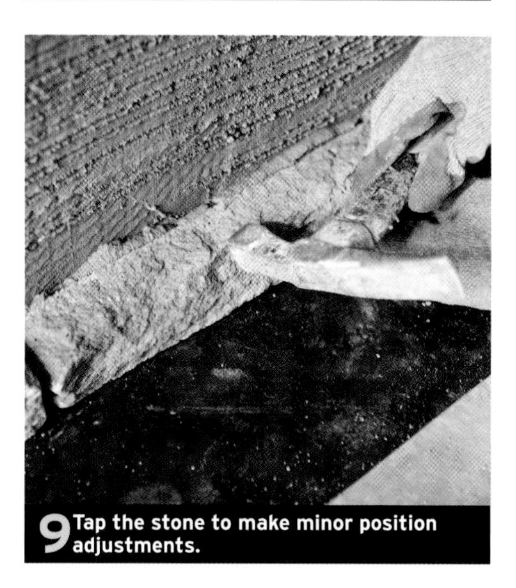

9 Tap the stone to make minor position adjustments.

10 Continue the next row with stones that match the corner stone thickness.

11 Use a brick hammer to break stones to fit, as needed.

A MATTER OF AESTHETICS

While understanding the mechanics of laying stonework is crucial, you also need to think about what makes stonework aesthetically pleasing. Luckily, 90 percent of what makes professional stonework rise above the rest can be achieved by following a few simple practices. To begin with, take the time to level the stones as you set them. This gives stonework an ordered and substantial appearance. If the stone you are setting is irregular, level the top. Typically, the top is the most visible surface, and a level top makes it easier to set the next stone level as well. Use shims (chips) to help stones stay in place while the mortar sets ❶.

While an ordered appearance is good, patterns such as long straight lines and running joints that draw the eye are visually unappealing. To avoid these patterns, do what's called "crossing" or staggering the joints. Don't align vertical joints, and periodically break up horizontal joints with larger stones ❷. >> >> >>

1 Level stones to give the stonework a more ordered and substantial appearance.

2 Cross, or stagger, horizontal joints to avoid long running joints that draw attention and look unnatural.

TRADE SECRET

To hide an unfinished end, set it so it faces a nearby wall; it will rarely be seen from that direction.

A MATTER OF AESTHETICS (CONTINUED)

Cultured stone breaks in a predictable manner when struck by a brick hammer. This is a big advantage when filling a spot for which you can't find the right-size stone. However, the broken end looks distinctively manufactured ❸, so avoid setting stone where these ends are visible.

Think through transitions to avoid gaps or overly wide joints. For example, where the face stone meets the underside of the flagstone hearth, allow for mortar by setting the top edge of the face stone up 5⁄8 in. to 3⁄4 in. ❹. The raised edge will ensure a tight joint under the lip of the flagstone. Bevel a small amount of mortar behind the top of the stones to hold them in place until you set the hearth ❺.

3 **Hide broken ends that show the cultured-stone aggregate.**

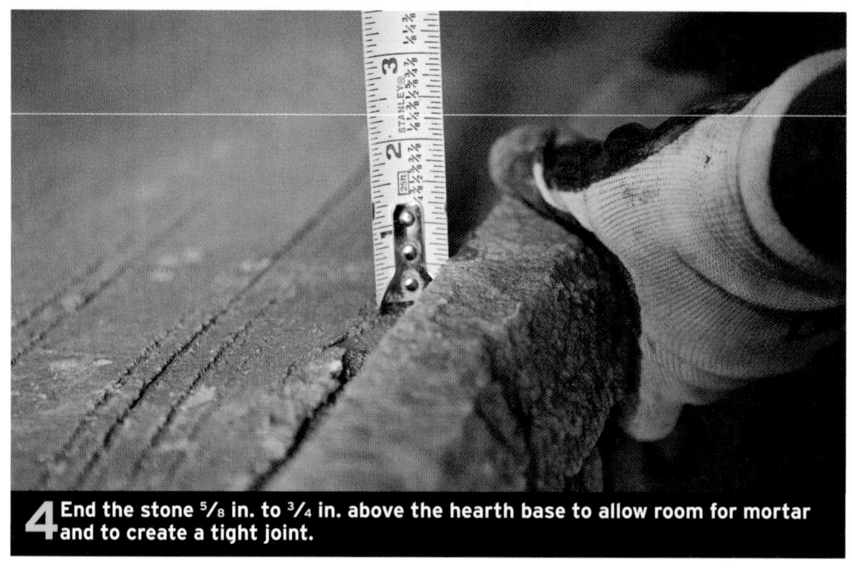

4 **End the stone 5⁄8 in. to 3⁄4 in. above the hearth base to allow room for mortar and to create a tight joint.**

5 **Bevel mortar behind the top course to secure the stones until the hearth is laid.**

SHAPING AND SETTING HEARTHSTONES

Laying a hearth employs essential stoneworking skills that can also be used for setting patio stones, capstones, and treads for steps.

Begin by shaping and setting the corner stones. First, select a stone that has a corner close to 90 degrees ❶. Place the framing square on the stone and mark a 90-degree cut line, wasting as little stone as possible. Use a chisel and hammer to remove the stone on the waste side of the line ❷. Make any small adjustments with the edge of the brick hammer as needed.

See "Dress a Corner Stone with Hammer and Chisel," p. 25.

On this fireplace, as on many, columns frame the sides. To fit the first stone around the column, shape the outside, or leading edge; it should fit the hearth with a 1-in. overhang. Then cut the back of the stone to fit snugly against the wall.

To notch the stone around the column, set the stone in place to the inside of the column and mark the depth of the column. Then measure the width of the column and transfer that measurement to the stone. Use a straightedge to draw cut lines for the required notch, and draw an X to indicate material to be removed ❸. Then, use a diamond blade to cut out the notch; cut as close to the inside corner as possible without overcutting ❹. A sharp rap with a brick hammer will free the piece and enable you to clean up the inside corner ❺. Before mixing the mortar, dry-fit the notch and corner; continue shaping the stone until you are satisfied. Then repeat the process at the other end of the hearth.

For setting hearthstones, use the same mortar recipe as for the wall stones (see p. 209). Use less water, however, to make a thicker, less soupy, mixture. You want the stones to be secure after you set them, not floating. >> >> >>

1 Select a stone with one corner already close to square. Then use a square to mark a cut line that wastes as little stone as possible.

2 Shape the stone as close to the line as possible. For thicker stones, use a chisel and hammer.

TRADE SECRET

To make quick work of shaping a corner stone, cut the angle with a saw or grinder fitted with a diamond stone-cutting blade. Then use a brick hammer to finish, or texture, the cut by chipping off the sharp corners.

SHAPING AND SETTING HEARTHSTONES (CONTINUED)

To set hearthstones in a wet mortar mix, apply mortar to the area the stone will cover. Then, with a trowel, channel the mortar ❻ so that the mortar has someplace to go as you set the stone. Without providing these channels, excess mortar would squeeze out from under the stone and make a mess.

Be sure the finished hearth will clear the insert face and any door or vent handles that protrude. If you have inches to spare, a quick visual check will suffice. We had no margin for error, so we used two levels to be sure the top of the stone would be below the fireplace face ❼.

Once the corner stones are set, use a long mason level to align the front edges and top faces ❽. Set stones to fill between the outboard corners and check alignment with a level ❾. Finally, set any remaining filler stones against the back wall, using a level to align the tops in the same way.

➔ **See "Fitting Stones," p. 110.**

After you've finished the hearth, you can set the stones above it immediately, as long as you don't touch the hearthstones. If it's too awkward to work without touching the hearthstones, let them set for 24 hours before continuing. You can grout the hearth after waiting 24 hours.

➔ **See "Dry-Mix Grouting," p. 114.**

Bottom to Top, or Top to Bottom?

Fireplace stonework is typically laid from bottom to top. However, for a fireplace large enough to require scaffolding, you may want to install the hearth last so falling debris doesn't knock freshly set hearthstones loose. To do this, measure the thickness of your hearthstones and add 1/2 in. for setting. Mark this height on the wall, and begin setting the stone veneer above it.

3 Place the stone next to the column and against the wall; mark the column's width and thickness on the stone. Mark waste with an X.

6 Use a trowel to form short channels in the mortar; they allow the stone to be set without causing excessive squeeze-out.

9 As you set stones between the corners, align the tops and front edges with a level.

4 Cut along the lines with a diamond blade. Be careful not to overcut.

5 Tap the corner to free the waste and clean the notch with a brick hammer.

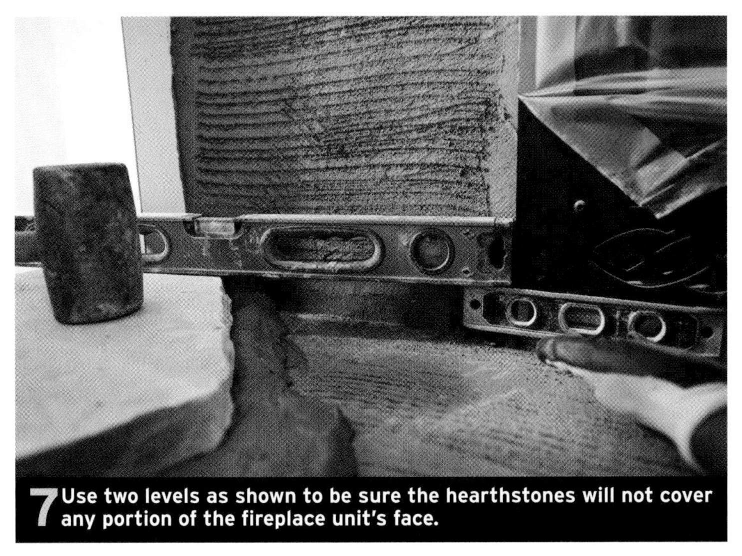

7 Use two levels as shown to be sure the hearthstones will not cover any portion of the fireplace unit's face.

8 Align the corners' front edges with a long mason level.

TRADE SECRET

Wet the scratch coat before applying the mortar. This prevents the cured cement from sucking water out of the new mortar and allows for a better bond.

SETTING STONES AROUND THE FIREPLACE

Set the courses around the fireplace insert, for the most part, in the same way as the stone below the hearth. Start by masking the hearthstones 2 in. from the wall to avoid staining them with wet mortar ❶. As before, start at the bottom by tilting the stone into place and holding it for a few seconds. With the narrow sides of a fireplace surround, it's more challenging to stagger joints, and you may need to cut or break more stones ❷. Again, try to avoid placing broken ends, with raw aggregate, where they can easily be seen. Position rough ends toward columns or a nearby wall.

Setting the keystone

For the arch, take the time to select a handsome keystone. You can buy a manufactured keystone, but a natural stone has more character and surface texture.

In self-supporting arches, the keystone is a critical structural element. It keeps the adjacent stones that form the arch (called soldiers) from falling. Even if the keystone is decorative, it should look substantial. With the rough keystone positioned above the fireplace, use a pencil to mark the shape you want. If your fireplace insert has a flange on the arch (as here), you can rest the stone there ❸. If your insert has no flange, have a helper hold the stone in place while you draw the shape. Double-check the shape's dimensions with a tape measure.

Use a diamond blade on a grinder or circular saw to cut the keystone rather than trying to shape it by chipping. You can add a rough edge with a brick hammer after it's sawn if you want that look. Position the stone again and use a level to check that the top and bottom cuts are parallel ❹. This may seem like a lot of fitting, but the keystone is a visually important element and you will notice if it is not shaped carefully. With that in mind, take a moment after setting the keystone to step back and see if it's exactly where you want it ❺.

Continue the arch on both sides with soldier stones. Size these stones shorter than the keystone to accentuate the latter. It's okay to vary the width of the soldier stones, as long as they are rectangular. With mortar, butter and set the keystone and soldier stones in the same manner as the coursework ❻. At the end of the soldier course, continue with normal coursework ❼.

1 **Mask the hearthstones. Then set the stone above the hearth in the same manner as you set the stones below the hearth.**

2 **Cross, or stagger, the joints between stones at random intervals.**

3 Position the keystone and draw the wedge shape directly on the stone.

4 Use a level to ensure that the top and bottom edges of the keystone are parallel. If necessary, draw a new cut line and trim.

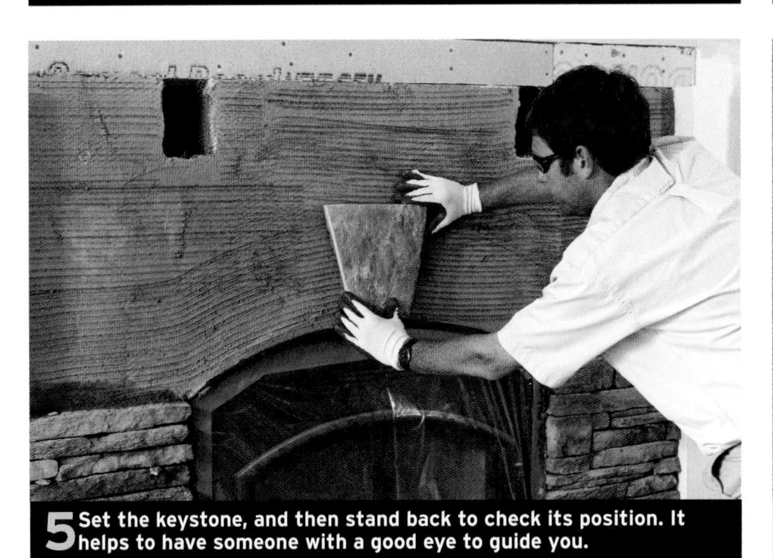

5 Set the keystone, and then stand back to check its position. It helps to have someone with a good eye to guide you.

6 Arrange and set the soldier stones to your own liking. It's acceptable to vary the widths, but they should all be the same length.

7 Continue courses up and around the soldier stones. Take the time to fit stones above the arch and regain a level course progression.

TRADE SECRET

Because the soldier stones should be the same length, you can cut them all at once, or "gang-cut" them. Lay the stones on the ground so that they align at one end. Measure and mark a cut line across the ragged end, and saw all the stones at the same time.

Cut the soldier stones slightly shorter than the keystone.

SETTING THE CORBELS AND MANTEL

Many stone mantels are made from one long stone with a natural, or "live," edge. Use two corbels, tied into the framing, to support it. Pick your mantel and corbels from the same stoneyard; choose them to match each other and the manufactured stone.

Install the corbels first

Install the corbels and mantel before setting the surrounding stones so you have room to work and won't loosen just-set stones. For this project, the contractor provided pockets in the framing and we cut the corbels to fit. If you're working with a contractor, it's important to plan ahead so there are no surprises. A heavy stone mantel like this one requires different framing inside the wall than a lighter, wooden mantel.

Sizing a mantel with a live edge is a little tricky. Use a straightedge to locate the back cut, choosing a line that takes full advantage of the stone's live edge and makes the best use of the stone ❶. Then locate and size the corbels so they have the proper setback from the mantel's edge and sit properly in the wall pocket ❷.

Cut the mantel

After you're happy with how the corbels will fit and you've double-checked that your mantel is sized properly, cut the mantel along the back edge. Do not try to make a cut this long and deep with a grinder. A circular saw with a diamond blade will work, but the best choice is to rent a gas-powered chopsaw fitted with a diamond blade ❸. It makes quick work of this big cut because the blade is large enough to cut through the entire thickness of the stone; there is no need to make a partial cut and then try to break the stone along the cut. That would be too risky with a stone this large and expensive.

➔ **See "Cutting Tools," pp. 38-39.**

1 Locate the mantel's back edge where it will best take advantage of the stone's width.

2 Dry-fit the corbels, making adjustments to the pockets or to the stones as needed.

TRADE SECRET

Extra framing is needed for a stone mantel for two reasons. First, the mantel and corbels need to be adequately supported. Second, the corbels require strong wall pockets into which they can be set. The last thing you want is for the corbels to tip out of their pockets and send the mantel crashing to the hearth!

Secure the corbels

To secure the corbels, spread a generous amount of mortar inside the pocket on all four sides; insert the corbel and shim it into place. Pack as much mortar into the hole as you can and use as many shims as needed to ensure the corbels are immobile. Once they are secure, finish setting manufactured stones around the corbels, but do not lay the last couple of courses of stone under the mantel until after it has been installed ❹. Now is a good time to scrape out the excess mortar and give the mortar around the corbels time to set. Allow at least one day before setting the mantel.

Set the mantel

When you are ready to set the mantel, spread mortar on top of the corbels and on the surface behind the mantel. It can take several people, one person on each end and two in the middle, to lift and set a mantel into place ❺. (If you don't have enough helpers to comfortably lift the stone, you can rent a lift from your local equipment rental store. Be sure to check that the lift is rated to handle the weight of your stone.) Use a rubber mallet to tap the stone into the mortar bed. While everyone is still there, stand back and make sure the mantel is centered. If you need to slide it a little one way or the other, now is the time to do it.

3 **Cut the mantel last. Use a power saw with a diamond blade to make the cut in one pass.**

4 **Set the corbels and surrounding stone. Leave out the last two rows until after the mantel is in position.**

5 **Lift the mantel in place while there's plenty of muscle on hand, and watch the fingers!**

FIRE PIT WITH SEAT WALLS

BUILDING A FIRE PIT WITH SEAT walls is a great way to turn an unused portion of your backyard into a place for family and friends to enjoy. The fire pit described in this chapter is a medium-sized project that will take 3 to 5 days for a mason with a helper to complete and at least twice that long for a novice. A fire pit with seat walls is also an excellent complement to an outdoor patio, and the skills learned from one project are readily applied to the other.

For this fire pit, we installed a mortared footing on a gravel base to allow for drainage and to simplify construction. For the benches, we poured earth-formed footings. Both types of footings, combined with mortared stonework, are well suited for walls around garden beds, small retaining walls, and property partitions.

PRACTICAL CONSIDERATIONS

If you are going to be burning a lot of yard debris, such as leaves and large branches, consider a large fire pit, with a 4-ft.-dia. or larger opening. For a smaller recreational fire pit, a 30-in. to 36-in. inside diameter is a good size. For any fire pit, take into account that the overall diameter will be 2 ft. larger than the inside opening (1 ft. on each side).

Also consider whether to build a fire pit with raised walls or flush to the ground. Raised fire pits allow people to sit on the walls or rest their feet on the edge when sitting in chairs. Raised walls also prevent hot ashes from blowing in the wind and shield you from some of the heat if the fire gets too hot. However, a fire pit flush to the ground is slightly warmer, can feel more intimate, and is good for people who prefer small fires (see the photo at right).

You will need to incorporate fire clay into the mortar when building the walls and cap of the fire pit to keep the mortar from cracking under intense heat. Fire clay, sold at most masonry supply stores, is a dry ingredient and needs to be added to the mortar mix prior to adding water.

When choosing the location of your fire pit, place it well away from the house to prevent smoke and sparks from blowing indoors. Similarly, put the pit well away from trees or shrubbery to reduce the danger of an accidental fire.

The owners of this fire pit **wanted a place to drink a glass of wine after dinner and enjoy the view. The fire pit did not need to be large.**

WHAT YOU'LL NEED

- 2 tons fieldstone
- 1 cu. yd. masonry sand
- 15 bags concrete mix
- 8 bags portland cement
- 2 50-lb. bags fire clay
- Wheelbarrow
- Mortar mixer (optional)
- Mattock or ax to remove roots
- Masonry tools
- Hand tamper

Two curved walls provide **permanent seating around the fire pit, while the open area allows the owners to add chairs and blankets for group settings.**

BEFORE YOU BEGIN

Look at the fire-pit layout options below and spend time thinking about the plan to be sure it will fit your needs. Build a fire in the spot you've chosen, and set up some chairs or buckets to sit on. Spend a few nights tweaking the arrangement to be sure you're happy with it; once it's set in stone, you'll have to live with it or spend considerable effort ripping it out.

Another option is to create the fire pit first, build a fire, and then try various seating arrangements. For the project shown here, we chose a compact design with two benches (or seat walls) around a smallish, 54-in.-dia. walled fire pit. If you want to use the fire pit before you build the walls, protect surrounding grass from the work with plastic (or plan on reseeding when the stonework is complete).

FIRE-PIT OPTIONS

Three common fire-pit layout options allow you to configure the space depending on how you will use it.

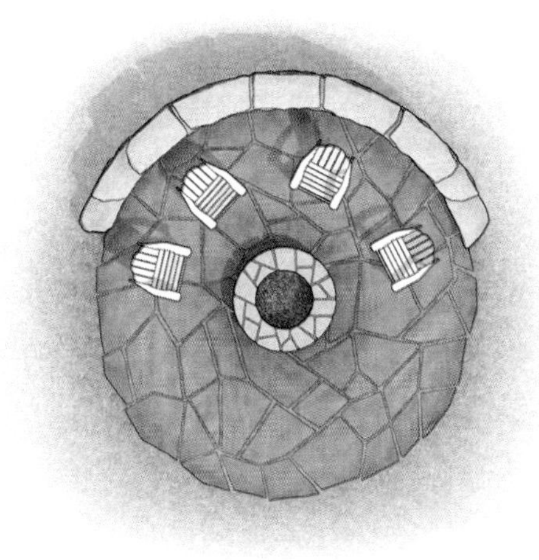

1. Fire pit with patio and a seat wall behind movable chairs. This design offers the most flexibility. Surrounding the fire pit is an open patio or lawn space where garden chairs can be freely moved depending on the situation. A seat wall, located behind the open area, defines the perimeter of the space and is a comfortable space for people to gather during parties and warm evenings.

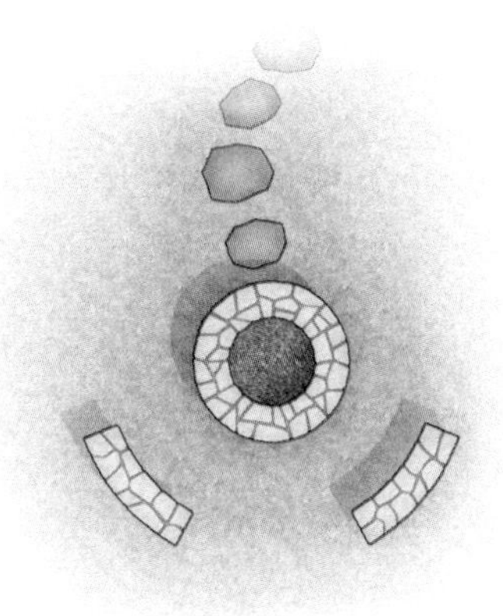

2. Fire pit with seat-wall benches. This is an intimate design with benches close enough to feel the heat from the fire (less than 4 ft.). It's ideal for a small yard and for people who want a place to sit but who don't want the hassle and clutter of lawn furniture.

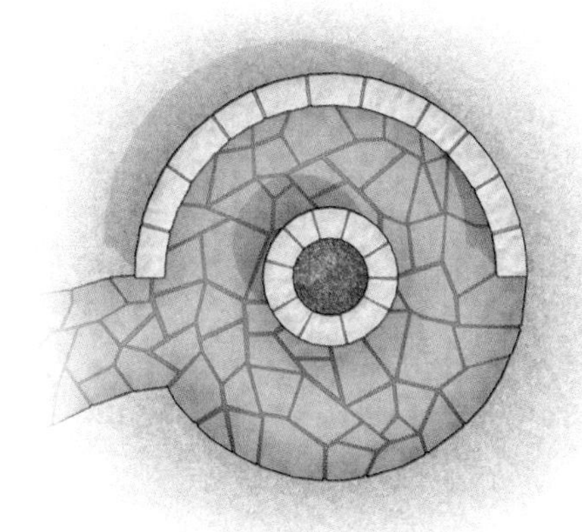

3. Fire pit with continuous seat wall. This design offers some of the benefits of both flexible and intimate designs. Here, the semicircular wall is close enough to the fire pit to feel the heat but far enough away so people can gather and enjoy the fire as a backdrop.

LAYOUT AND EXCAVATION

For this fire pit, we used compacted gravel as the base rather than a poured concrete footing. A gravel footing allows water to drain quickly and is easier to install than concrete.

Layout

A garden hose can help you visualize where to put the fire pit; you can move the hose around as much as you want. A hose can also help you understand how big the finished fire pit will be ❶. It's not, however, a substitute for drawing the plan on paper or properly measuring and marking a layout on the ground.

To lay out a perfect circle, measure the pit's diameter and divide by two to find the radius. Mark the pit's center ❷ and set the end of the tape on it. Rotate the tape around the center, and mark the circumference at 10-in. to 12-in. intervals. ❸. Connect the dots with a continuous line around the circle ❹.

Excavate and add gravel

Use a spade to cut along the circumference, and then dig the footing to a depth of 8 in. to 12 in. or deeper until you find undisturbed soil ❺. When digging, take care to cut clean, vertical sidewalls ❻. Later, you'll use the edge of the hole to set the perimeter stones; any irregularities will transfer to the stonework, so dig a perfect circle. Also create a flat base, without bumps and without depressions that can hold water ❼. With a hand tamper, thoroughly compact the pit base; pay special attention to the area near the circumference upon which the wall will be built ❽. Add 3 in. of gravel and spread it evenly with a rake or flat shovel ❾. As with the soil base, it's important to create a flat gravel surface without dips or bumps. Use a level to check for flatness and for any desired slope; then compact the gravel thoroughly ❿.

1 Use a garden hose to help you determine the size and placement of the fire pit. Here, the new pit will replace an existing pit.

4 Connect the marks to form a circle. Mark the line carefully; it will become the fire-pit circumference.

7 Create a flat floor without depressions, which might hold water. Pitch the floor slightly in the direction of the surrounding grade.

8 Compact the base soil, especially near the circumference where most of the weight will be.

2 Mark the fire pit's outside wall location using the hose as a reference; then divide the diameter in half and mark the center.

3 Using the center mark, make a series of marks along the circumference at intervals of 10 in. to 12 in.

5 Excavate using the line as a reference. For more accurate results, consistently dig to the outside edge of the line.

6 Use a sharp spade to dig vertical walls. Square the walls to the base to allow stone to set snugly to the circumference.

9 Add 3 in. of gravel, and use a rake or shovel to smooth it.

10 Compact the gravel to complete the fire-pit bed. As with compacting the soil, be extra vigorous near the circumference.

LAYING THE FIRST ROW

Fire pits are not likely to fail. The circular design is inherently strong, and there is no force on the fire-pit wall from soil being retained. By carefully locating the fire pit and properly preparing its base, drainage and other water-related issues are easily dealt with. For these reasons there is no need to incorporate a large footing or rebar.

To avoid extra movement while laying the first course (and risk stepping on and collapsing the carefully cut sidewall), shape a decent pile of stone and put it within easy reach ❶.

For a jointed fire-pit wall like the one shown here, mix mortar using one part portland cement, one part masonry sand, and two parts fire clay. To avoid staining the stones, do not mix the mortar too wet. ❷. For a mortared wall in the dry-stack style, mix a wetter, but not soupy, mortar. Spread the mortar bed 3 in. deep and 12 in. wide ❸. If you dug the sidewall of the hole with care, simply butt the end of the tape against the soil to measure for the mortar bed ❹.

➔ **See "Hand-Mixing Mortar," p. 62.**

Lay the first stone against the earthen sidewall, and pack mortar behind the stone to stabilize it ❺. When laying stones for the first course, try to use stones with tops and bottoms that are parallel, which will make setting a stable foundation easier. Measure 12 in. from the outside face and set the face of the inside stone ❻. Then pack mortar behind the inside stone to stabilize it.

Continue setting the first course of stones using the side-wall of the pit excavation as the outside reference and measuring back 12 in. to set the inside stones ❼. Level the tops of the stones as you set them ❽, and pack mortar underneath to keep them level. When the first course is set, fill between the stones with shards and rubble left over from shaping (set them in the mortar bed rather than just dumping them in) ❾. Fill the rest of the space with mortar using your trowel and fingers to pack the mortar into the crevices ❿. Add mortar to fill to the top of the stones, and then pack it down with a trowel ⓫. Continue adding and packing mortar until it doesn't yield to moderate pressure. To complete the first course, pack mortar between the joints with a pointing trowel ⓬. >> >> >>

TRADE SECRET

To attain the right mortar consistency, mix the dry ingredients first. Then add small amounts of water until you can form the mixture into a ball with one hand and the ball stays together when tossed in the air.

LET THE WATER OUT

Set a drainpipe between the first and second course to prevent your fire pit from becoming a water feature. Make sure the pipe has a minimum 1 in. diameter and is about as long as the wall is wide. Set the pipe just above finish grade and pitch it to the outside for proper drainage. So the pipe doesn't interfere with the coursework, set the pipe in a joint where four corners intersect. Grout around the pipe after you set it and continue stacking the stones.

1 Stage wall stones within easy reach. The fewer times a stone is handled, the more efficient you will be with your work.

2 Mix mortar to the consistency of dry cottage cheese. The mortar should hold together when pressed but not be wet.

3 Spread a layer of mortar 3 in. deep around the outer rim where you will begin laying stones.

4 Limit the mortar to a 12-in.-wide band. Keep the fire-pit center free of mortar to allow water to drain.

5 Place the first stone to the outside against the earthen wall. Select a stone with parallel top and bottom, or shape one that way.

6 Wall stones, measured face to face, should be 12 in. apart. Push mortar behind the stones to stabilize them.

LAYING THE FIRST ROW (CONTINUED)

7 Continue setting wall stones around the fire pit. Measure often to maintain the 12-in. wall depth.

8 Level the tops of stones as you set them. If a stone needs to be adjusted, pack mortar behind and under it until it's level.

9 Fill the wall interior with rubble and shards left over from shaping stones.

10 Pack mortar between the stones and rubble. Use a trowel and your fingers to work the mortar into all the crevices.

11 Fill the wall with mortar flush to the top of the first course of stones. Pack the mortar until it is firm and smooth the top.

12 Pack the joints with mortar using a pointing trowel. Fill the joints until they are slightly mounded.

CONTINUING THE COURSEWORK

Continue building the circular fire-pit wall by spreading a 1-in. layer of mortar where you'll set stones. Use your hand and trowel together to build up mortar at the edge ❶. Align the face of each stone with the course below, and gently tap it into the mortar bed ❷. Vary the stone heights to break up long-running horizontal joints ❸ and create a more pleasing overall pattern.

To set the stones on the fire pit's inside face, use your hand and trowel to build up a 1-in.-thick mortar bed along the inside edge ❹. Set a stone with the face aligned with the course below ❺. Then level the top and pack mortar behind it ❻. Use a pointing trowel to pack mortar into the joints until it is flush with the face ❼. Sometimes it's more efficient to lay a long course of stones and then pack mortar (or grout) around the stones all at once, instead of working each stone individually. Make sure you grout all around the stones of one course before beginning the next. Finally scrape the joints flush after they dry—but before they harden—using a pointing trowel ❽. Brush away the excess mortar with a whisk broom ❾.

As you work, measure across the stones frequently to make sure you're maintaining a 12-in. wall thickness ❿. Also check the wall, inside and out, with a level to be sure it stays vertical ⓫. The inside and outside rings do not need to stay level across the top while laying courses until you reach the top. As with the first course, fill the inside of the wall with rubble and pack the crevices with mortar ⓬. It's important to fill all the voids, so keep adding mortar and pressing it down until it is firm. Completely fill between the stones and then smooth the top of each course. >> >> >>

1 Spread a 1-in.-thick bed of mortar for the first few stones of the second course.

2 Align and set second-course stones to the course below.

3 Vary the height of the stones to avoid long-running horizontal joints that will draw attention to themselves.

CONTINUING THE COURSEWORK (CONTINUED)

4 Set the inside ring of stones the same as the outside ring. Build up an edge by pressing the mortar between your hand and trowel.

5 Select or shape stones to match the inside curve. Cross the vertical joints whenever possible.

6 Pack mortar behind the stone to level its top.

7 Pack mortar into the joints with a pointing trowel. (Some masons set a whole row of stone and then pack all the joints at once.)

8 Scrape the excess mortar from between the joints after completing each row.

9 Brush excess mortar from the surface. If it stains the concrete while brushing, it is too wet and you need to let it dry longer.

10 Check the wall width frequently for consistency. With irregular stones, it is easy to overlook a 1-in. discrepancy.

11 Check the outside of the wall for plumb and straightness. Having plumb and even walls greatly improves the overall look.

12 Fill between each course of stones with rubble and mortar. Work the mortar between the stones to increase wall strength.

FINISHING THE COURSEWORK

1 **Establish the top-of-wall height by setting a stone 15 in. above grade. If the grade slopes, measure in the middle of the slope.**

For a 19-in.-high wall (as for this project), the regular coursework ends at 15 in. ❶. This leaves room for mortar and 3-in.-thick capstones. To level the top of the coursework all the way around, pick one spot that is the correct elevation. Then set the opposite-side stone to the correct height ❷. Repeat this process, setting a pair of opposing stones 90 degrees from the first pair ❸. Use these four stones as a guide to set adjacent stones to the correct elevation ❹. As you work, level across the wall to be sure that both the inside and outside faces are at the same elevation ❺. For stones that rest slightly high, it is better to mark and cut the stone to the proper elevation rather than set it deep and reduce the joint size ❻. Fill the inside of the wall with mortar, and pack and smooth the top ❼.

2 **Level across the wall and set another stone at the same elevation. Shape the stone rather than adjust the joint size.**

3 **Use a level to set two more stones at the proper elevation, 90 degrees from the first pair.**

4 **Set adjacent stones by leveling across the tops until the entire outside ring is set.**

5 Level across the wall's width to set the inside stones to the same elevation.

6 Use a level to mark a straight line if you need to trim the top of a stone.

7 Pack mortar between the stones and smooth the top of the wall. You are now ready to set the cap.

MAKING CURVED STONEWORK

Cut a slight curve onto the face of each stone as you set them around the circle. To do this, take the sharp edge of a brick hammer and chip away the face of the stone in a downward motion on the left and right side of the stone. Some stones will have a natural curve to them, which can save you a lot of work. Train your eye to look for these stones and take advantage of them whenever possible.

SHAPING CURVED CAPSTONES

Shaping and setting capstones in a curve is essentially the same as setting other stones except there is more fitting and shaping. Because the process takes more time, shape stones for the entire cap and dry-fit them before spreading a mortar bed. As with the wall stones, take advantage of stones that already have some curve to them ❶. For stones that don't have a ready curve but are generally the right size, position the stone to remove the least amount of waste but still fit the space. Then sight from above and mark the cut line ❷. For longer stones that cover more of the curve and make it harder to sight from above, trace the curve of the wall on the bottom of the stone ❸. Then allow for the overhang when shaping it ❹.

1 Position capstones and mark the curved cut line; allow for the overhang (in this case, 2 in.).

2 Sight the stones from above to mark where to remove material. Position the stone to remove as little material as possible.

3 Mark the underside of stones that are long and difficult to see from above. Here, a stone chip is used to mark the curve.

4 Shape the stone's edge 2 in. from the line to allow for the overhang. You may have to shape the edge from both the top and bottom.

SETTING CAPSTONES

Once all the capstones are shaped, lay a 1-in. mortar bed and form an edge with the back of your hand ❶. Stabilize and level the stones across the wall by pushing mortar beneath them. Then set each stone level with adjacent stones by placing a level across the top and tapping the stone into the mortar ❷. If you're concerned about chipping or scratching the caps, use a rubber mallet to set them. As you progress, frequently measure across the wall to check for consistent cap width. Remember to hold the tape so that, if extended, it would intersect the center of the pit; otherwise, your measurements will be off ❸. Also, after setting each stone, check the overhang ❹. It is very easy to wander off the line, especially when setting curved stones.

1 For each capstone, spread a 1-in. layer of mortar and build up the outside edge slightly with your trowel and hand.

2 Set the stone by tapping it into the mortar. Align it with the tops of adjacent stones and level it.

3 Measure the wall cap width frequently as you set capstones.

4 Check the overhang of each stone before setting the next one. Also make sure the corners align to avoid a jagged appearance.

LAYING OUT THE SEAT WALLS

Most of the techniques used to shape and set wall stones and capstones for a fire pit are also used to build seat walls. The seat walls are curved, mortared walls, 12 in. wide and 16 in. to 18 in. high, capped with 3-in.-thick fieldstone or flagstone (we used the latter for our project). That said, there are some key differences that involve a few extra steps with layout, installing the footing, and finishing the ends of the walls.

Orienting walls on a radius curve

Lay out the curved seat walls using the fire pit's center to establish a radius. Generally, 3 ft. to 5 ft. between the seat walls and the fire-pit wall allows enough room for people to move around comfortably—and yet still feel the warmth of the fire. However, the exact location of the seat walls really comes down to personal preference. As mentioned earlier, a good way to decide where to put seating is to grab something to sit on ❶, and move around until you find a spot where you feel comfortable. Then mark the spot, and lay out seat walls to match the curve of the fire pit wall.

Drive a tall stake into the ground at the center of the fire pit ❷. Then drive a nail into the stake, make a loop at the end of a length of twine, and slip it over the nail ❸. You can also use the string of a plumb bob. Pull the string to the seat-wall location ❹, and hold it taut. Then, holding a marking can at the end of the string, rotate around the stake and mark an arc ❺. This line represents the front of the seat wall ❻. Mark a second arc for the back of the seat wall 12 in. from the first ❼. Last, carefully square the ends at the length you desire ❽.

1 Sit on a bucket to verify the seat wall location. Even if the location is already established, it won't hurt to confirm it's what you want.

4 Pull the string to the seat-wall location and mark the ground with marking spray.

PULL A RADIUS TO LAY OUT SEAT WALLS

Although the seat walls have a longer radius than the inside circle of the fire pit, the curve is still established by rotating from the fire pit's center.

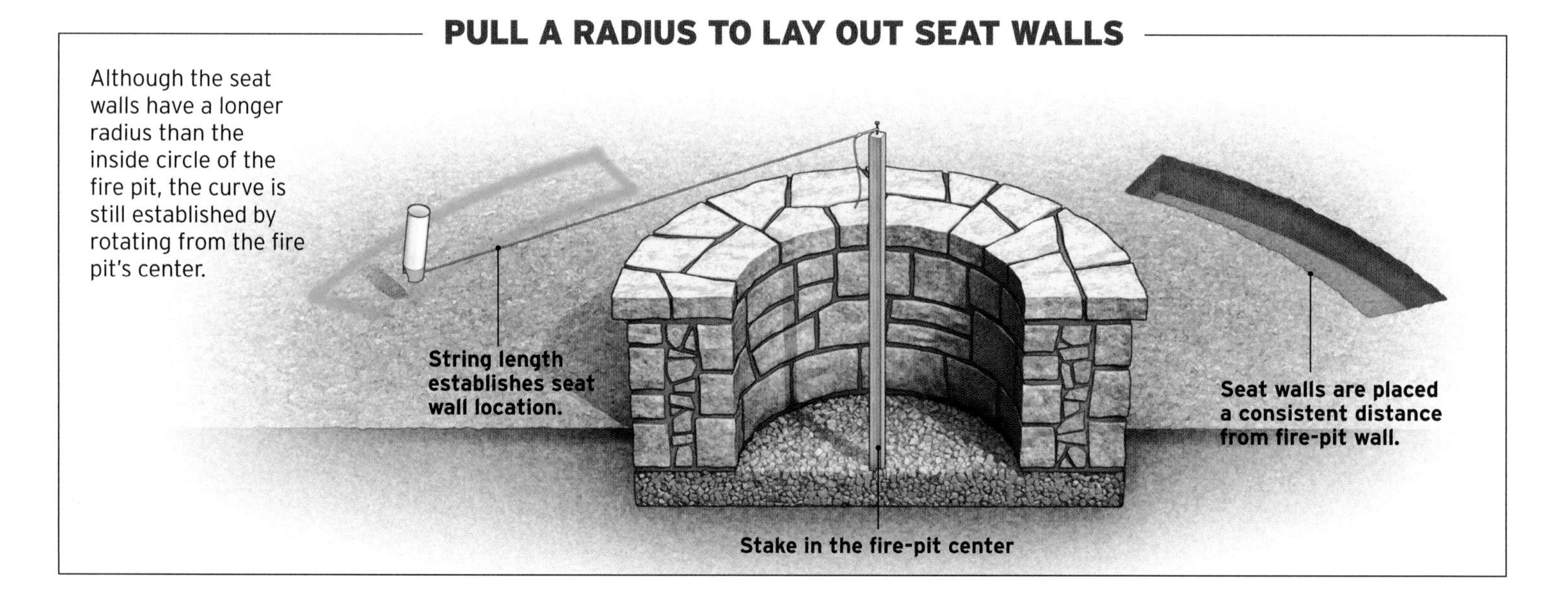

2 Drive a stake into the fire pit's center. Make sure the stake is secure and extends above the fire-pit wall.

3 Drive a nail into the center of the stake and attach string to it. Avoid tying the string to the nail; it will need to rotate freely.

5 Hold the marking spray nozzle and twine end together, with the twine taut, and rotate (swing) while spraying.

6 Mark an arc equal to the wall's length. Check that the line doesn't wiggle. Any irregularities will transmit to the wall.

7 Measure 12 in. farther away and mark another arc. This represents the seat wall's outside edge.

8 Square the wall's ends. Repeat this process for each wall in your project. For this project, we outlined two walls.

INSTALLING THE FOOTING

An earth-formed concrete footing adds stability to a small wall that might otherwise shift or tilt. There usually isn't a need to add rebar. If you locate a seat wall adjacent to a driveway, where it may get bumped, consider adding rebar with vertical ties.

Carefully dig the footing trench following the marked lines as closely as possible ❶. Dig to a depth of 8 in. and then square the sides and level the bottom of the trench. Use a tamper to compact the soil ❷. Add 3 in. of gravel and use the tamper again to compact the gravel thoroughly ❸. Mix and add concrete until it fills the trench to 1 in. below grade (2 to 3 bags of premixed concrete). The surface does not need to be pretty, so you can flatten it with the back of a square shovel. However, it does need to be level in two directions ❹. Let the footing cure for 24 hours before building the wall.

1 **Dig the footing trench using the line as a guide.**

2 **Compact the footing base with a hand tamper. Keep the walls as vertical as possible and square at the base.**

3 **Evenly spread 3 in. of gravel in the footing trench and compact thoroughly.**

4 **Fill the trench with concrete to 1 in. below grade. Smooth the surface with a trowel or square shovel and check for level.**

BUILDING MORTARED WALLS

The easiest way to build a short, mortared wall is to build the ends first and then build the sides for each course. Spread a 2-in. layer of mortar for the end stones ❶. Set two corner stones the width of the wall apart ❷. Fill the space in between them with another stone, creating approximately 1-in. joints. Stabilize the stones by packing mortar behind them ❸. Tap the stones into the mortar to fully set them, and level their tops ❹.

Repeat these steps at the other end of the wall ❺. Then, lay the first course between the ends in the same manner as you laid the fire-pit walls (set the inner and outer stones and fill with rubble and mortar) ❻. As you build the wall, check that each corner remains straight and plumb ❼. Pack mortar in the joints for each course before you lay the next course. Scrape the joints flush once they have dried a bit but before they set up. At the top of the last course, 13 in. to 15 in. above grade, level the stones. Then pack the wall's interior with mortar and level the top ❽. >> >> >>

1 Spread mortar 2 in. thick on the footing for the end stones.

2 Use a measuring tape to set corner stones to the wall width. The tops should be level but do not have to be the same elevation.

3 Set a stone between the corner stones, sized to provide a 1-in. joint; secure all three stones by packing mortar behind them.

4 Tap the stones into the mortar bed to fully set them. Tap until the tops are level.

5 Set the seat wall's other end in the same manner. In this case, we used two stones to make up the corner, instead of three.

BUILDING MORTARED WALLS (CONTINUED)

6 Spread mortar and set the first course of stones between the end stones; try to create an even curve.

7 The eye is naturally drawn to the corners, so make them as straight and plumb as possible.

8 Pack the wall with mortar; then flatten and smooth the top before shaping the capstones.

WARNING

You or your guests may not be as sensitive as the princess who couldn't sleep on a pea, but sitting on even a small bump can get uncomfortable during a crisp spring night. Treat yourself right by aligning stones carefully and avoiding lumps, bumps, and dips on stone faces.

SETTING THE SEAT CAPSTONES

Setting capstones for a curved bench top is similar to setting capstones for a circular fire pit. Pay extra attention, however, to the surface of candidate stones before selecting them. Small protrusions, bumps, or dips that would otherwise go unnoticed are uncomfortable to sit on. When setting stones, aim for a smooth surface without height changes at joints.

Cut all the caps and dry fit them on the ground before mixing mortar. This makes it easier to make slight adjustments and will prevent the setting mortar from drying while you make the cuts. Try to arrange the stones so at least a couple will span across the top of the wall and avoid a running joint. End stones are good candidates for this ❶. Stones that have a diagonal side can make fitting easier. Set stones with opposing diagonals across from each other, and shift them slightly to change the overall width while keeping the joint width consistent ❷.

Use the same mortar recipe for setting caps as you did to build the walls, but add a little extra water. This recipe creates a less-porous mortar that helps prevent water from seeping into the wall and efflorescence from leaching out. Spread a 1-in. layer on the walls ❸ and set the caps perfectly level in two directions. ❹. Even a slight slope will be noticeable after sitting on it for a few minutes.

➔ **See "Preventing Efflorescence," p. 73.**

Once the capstones are set, use a pointing trowel to pack mortar into the joints, letting it mound up slightly on the edges ❺. Let the mortar dry until you can scrape it without leaving stains on the capstones—but before the mortar has set ❻. Scrape the mortar flush and sweep away the excess.

1 **Span the entire wall width with capstones at least a couple of times.**

2 **Create some diagonal joints. They break up the monotony of parallel joints.**

3 **Spread a 1-in.-thick mortar bed for the capstones.**

4 **Level the capstones in all directions.**

5 **Pack mortar into the joints with a pointing trowel.**

6 **Scrape the joints flush to the capstones.**

WALL REPAIR

It's common for old stonework to develop cracks and for stones to loosen over time due to freeze-and-thaw cycles and to the eventual breakdown of mortar. Making necessary repairs is not difficult, but there are some challenges. For one, the new grout should match the old grout. Similarly, if the old wall has lost stones along the way, you'll have to find new stones that match old ones. With a little time, patience, and the right tools, however, even a novice will find success.

Repairing wall cracks

Cracks in retaining walls and freestanding walls can develop for a number of reasons: The walls have old mortar that is falling apart, they have too much pressure exerted on them due to hydrostatic pressure, they have roots pushing into the back of the wall, or the wall has settled because of an inadequate footing. Sometimes the problem is a combination of both pressure and settling. If the wall is cracking for these reasons, often there is no way to address the fundamental problem without tearing out the wall and starting over. But the good news is that a patch may last many years. Given how easy it is to do, it is usually my first choice.

To repair cracks, use a hammer and small chisel to remove loose stones along the crack ❶. Make sure you remove all of the mortar, even if it is attached to the front of the stones. Don't, however, strike the mortar too hard; you don't want to loosen adjacent stones. If the mortar doesn't break loose with a few hammer taps, leave it and grout around it. The objective is to leave a wide enough gap to pack in an adequate amount of new mortar. Clean the stone edge where mortar was removed ❷.

Use a whisk broom or stiff brush to remove debris and dust particles ❸. Mix a batch of mortar and apply the grout, using the wet-grout method ❹ or dry-grout method ❺, whichever you prefer. Insert the grout until it's slightly proud (raised from the surface) of the stones. This will leave you room to scrape it after it sets up. Use a small trowel to spread the mortar in the joints if the crack is large ❻.

Check the grout after about an hour to see if it's ready to be pointed. If it is slightly hard, rake it flush with your pointing trowel. If the mortar is still wet, let it dry longer ❼. If you are matching a specific style of mortar joint, finish the joint accordingly.

➔ See "Dry-Mix Grouting," pp. 114-115.

➔ See "Using a Grout Bag," pp. 116-117.

>> >> >>

1 Remove cracked, crumbling, or decaying mortar with a hammer and chisel. Do not remove solid mortar.

4 Apply or pack mortar into the cracks. The more mortar you can pack in, the stronger the repair will be.

7 When the mortar dries, scrape it flush or shape it to match the existing joint style.

2 Use a hammer and chisel to clean the stone edges where mortar was removed.

3 Remove dirt and debris well back from the stone face. A stiff brush can help.

5 Fully pack large cracks with mortar. For large cracks, we use a dry-grout method (see pp. 114-115).

6 Finish the mortar slightly proud of the surface. For large cracks, use a small trowel for packing the mortar.

WALL REPAIR (CONTINUED)

Securing loose capstones

To repair loose capstones, first remove the stone even if you have to knock it a few times to free it ❶. Use a hammer and chisel to remove old bedding material or adjacent mortar ❷. Remove loose grout from the stone location as well as any on the stone itself. Continue removing loose stones until only solidly held stones remain.

Mix a batch of mortar, similar to flagstone-setting mortar (see recipe on p. 61), and spread a 1-in. layer on the wall under the location of the cap. Use your trowel to spread the mortar ❸. Set the cap on the mortar bed and tap it into place with a rubber mallet or the handle of your hammer ❹.

Use a pointing trowel to pack mortar into the joints on both the horizontal portion of the stone and the vertical. Leave the grout raised slightly above the stone until it sets up (about an hour). Then scrape it flush with the joints, or shape it like the joints you are trying to match ❺. After scraping the joints, use a whisk broom to sweep away any excess.

1 Remove loose stones and set them aside. If necessary, mark them to identify their original location.

4 Replace the capstone and tap it into place with a hammer handle or a rubber mallet.

Preventing cracks

To prevent cracks from developing in a new wall, pour an adequate-sized footing, build the wall thick, use an ample amount of gravel when backfilling, and provide adequate drainage. You also might consider removing small trees that are behind the wall so their roots do not eventually push it over.

Additionally, take the time to ensure you are using the proper mix ratios for the mortar and tightly pack all voids in the wall with mortar. If you do all of this correctly, your wall should last for many years to come.

2 Remove loose mortar, as well as previous bedding material.

3 Spread a new bed of mortar as you would for installing a new capstone.

5 Grout the joints as you would for a new capstone, packing the mortar proud of the joint and then scraping it flush.

GLOSSARY

ANTIFRACTURE MEMBRANE A material applied under flagstone that prevents cracks from transferring to the stone.

BACKERBOARD Also known as cement board, a combination of cement and reinforcing fibers formed into sheets. It can be attached to wood or steel studs to create a substrate for vertical stone or tile and attached horizontally to plywood for stone or tile floors, kitchen counters, and backsplashes. It can be used on the exterior of buildings as a base for exterior plaster (stucco) systems and sometimes as the finish system itself.

BATTER A slope on the outer face of a wall that angles in from bottom to top.

BATTER BOARDS Horizontal boards that are nailed at right angles to each other to stakes set beyond the corners of a structure for excavation. Strings are fastened to these boards indicating the exact corner of the structure.

BRICK A block of ceramic material used in masonry construction.

BROWN COAT A layer of mortar that is applied over a scratch coat to prepare the plaster or stucco base for the finish coat application.

BUILDING CODE A set of rules that specify the minimum acceptable level of safety for structures.

BUILDING DEPARTMENT Provides plan checking and building inspection services to ensure compliance with local building codes.

BUILDING INSPECTOR A person with the authority to control building work that is subject to building regulations.

CAPSTONE The top stone of a structure or wall.

CEMENT A building material made by grinding limestone and clay to a fine powder, which can be mixed with water and poured to set as a solid mass or used as an ingredient in making mortar or concrete.

CHAIR An object used to elevate rebar for concrete slabs and footings.

CONCRETE A composite construction material composed primarily of aggregate, cement, and water.

CONCRETE BLOCK A large rectangular brick used in construction. Concrete blocks are made from portland cement, aggregate, and sand.

CONTROL JOINT A planned crack that allows movement due to ground settlement or earthquakes.

CORBEL A projection jutting out from a wall to support a structure above it.

COURSE A continuous horizontal layer of similarly sized building material in a wall.

CROWN The top of a masonry chimney, usually finished with concrete made smooth by a trowel.

CURE The hardening of concrete or mortar.

DRAIN LINE A pipe used for moving water from one place to another.

DRAINAGE The means of removing surplus water or liquid waste.

DRAINAGE MAT A meshlike material installed behind masonry veneers that allows water to drain properly.

DRESS To surface and shape pieces of stone.

DRY GROUT Grout that has less moisture than wet grout. Primarily used for grouting with your hand, a trowel, and tuck pointer, as opposed to wet grout, which is applied with a grout bag.

DRY-STACK A type of stonework in which the mortar, if there is any, can't be seen.

EFFLORESCENCE A whitish, powdery deposit on the surface of rocks, block, or brick, which is formed as mineral-rich water rises to the surface through capillary action and then evaporates. Efflorescence usually consists of gypsum, salt, or calcite.

EROSION The process of wearing away by wind, water, or other natural agents.

EXCAVATE To dig out and remove soil.

FELT Also known as tar paper, a heavy-duty paper used in construction to prevent the ingress of moisture.

FILTER FABRIC A material that holds soil in place and prevents small particles from entering and clogging a drainage region.

FINES Finely crushed stone dust that is used in patio and walkway construction.

FINISH The outermost layer of masonry, such as stone or brick veneer or stucco.

FINISH COAT The final coating of plaster or stucco applied to walls.

FLAGSTONE A flat slab of stone used as a paving material.

FOOTING The supporting base or groundwork of a structure.

FOUNDATION The lowest support of a structure.

GRADE Ground surface; also a degree of inclination of a slope.

GRAVEL Small pieces of rock used in footings and foundations.

GROUT A mortar or paste for filling crevices, in particular the gaps between stone or brick.

HEARTH The floor of a fireplace, usually extending into a room and paved with brick, flagstone, or cement.

HYDROSTATIC PRESSURE Pressure exerted by water behind a retaining wall due to poor drainage.

JOINT A void between brick, stone, or concrete masonry units.

KEYSTONE A central wedge-shaped stone of an arch that locks its parts together.

KNOCKDOWN FINISH A drywall or stucco finishing style with a mottled texture, more textured than a simple flat finish.

LANDSCAPE FABRIC A loosely intertwined fabric placed on soil to inhibit weed growth.

LATH A mesh of metal that provides a base for plaster or stucco.

LEVEL A device that uses a bubble in a tube of liquid to establish a horizontal line or plane.

MANTEL The protruding shelf over a fireplace.

MANUFACTURED STONE Man-made replicas of natural stone used for veneers. Manufactured stone is cast in flexible molds, hand-dyed with iron oxide pigments, and made of lightweight aggregate materials. Also known as cultured stone.

MASONRY The building of structures from natural or man-made material, laid in and bound together by mortar.

MOISTURE BARRIER Also known as vapor barrier, used to refer to any material that prevents the ingress of moisture, especially on walls before masonry is applied.

MORTAR A mixture of lime or cement with sand and water, used as a bedding and adhesive between adjacent pieces of stone, brick, or other material in masonry construction.

NATURAL STONE A stone that occurs in nature, as distinguished from a man-made substitute, used for decorative purposes in construction.

PAVERS A molded rectangular block of clay or concrete used as a paving material.

PEBBLEDASH FINISH An exterior stucco finish containing crushed rock, large pebbles, or shells that are embedded in a stucco base.

PITCH The slope of a surface usually expressed as a ratio of vertical rise to horizontal run.

PLUMB Exactly vertical, or in a vertical or perpendicular line.

POINTING The process of repairing a mortar joint in a brick or stone wall.

POLYMERIC SAND A granular material placed between the joints or seams in brick, pavers, or stone in patios, walkways, and driveways.

PORTLAND CEMENT A basic ingredient in concrete, mortar, stucco, and most non-specialty grout.

QUARRY DUST Finely ground quarry stone used in setting flagstone, brick, or pavers.

REBAR A steel rod or bar with ridges, used for concrete reinforcement.

REINFORCED CONCRETE Concrete in which wire or metal bars are embedded to increase its tensile strength.

RETAINING WALL A wall that holds back earth on one side.

RISE The distance from a lower to a higher position, such as steps.

RISER The vertical part of a stair step.

SAND A loose granular material that results from the disintegration of rocks. It consists of particles smaller than gravel but coarser than silt and is used in mortar and concrete.

SAND FINISH Applied using a foam float that brings sand grains evenly to the surface.

SCAFFOLDING A temporary metal or wooden framework used to support workmen and materials during construction.

SCRATCH COAT The first coat of plaster or mortar, scratched to create a strong bond with the second coat.

SEAL To apply a waterproof coating.

SET To put into a stable position.

SETBACK A particular distance from a curb, property line, or structure within which structures are prohibited.

SLAB A flat, reinforced-concrete horizontal surface.

SLOPE A surface in which one end or side is at a higher level than another.

SOLDIER COURSE A course of upright bricks or stone with the narrow faces showing on the wall surface.

STONE DUST A finely crushed material used to fill the spaces between gravel and paving stones.

STRINGLINE Twine used in masonry to establish a straight line for a retaining wall or patio. It is also used to measure the slope of patios or walls and to make straight corners on columns and wall ends.

3-4-5 METHOD Way to check a corner's squareness.

TREAD The upper horizontal part of a step in a staircase.

TROWEL FINISH A smooth finish given to stucco or concrete with a trowel.

VENEER A masonry wall that consists of a single nonstructural external layer of masonry work, typically brick or stone, that is sometimes backed by an air space.

WALL TIES Metal strips or wires used to tie a masonry veneer to a wall.

WEEP HOLES Small holes in a wall located near lower ground surface that allow moisture drainage, preventing water pressure buildup.

WET GROUT Grout method using a grout bag to distribute mortar into joints.

WIRE MESH Wire in a gridlike shape used to reinforce concrete.

INDEX

Note: Bolded headings refer to projects.